Indigenous Futures and Learnings Taking Place

Singularizing progressive time binds pasts, presents, and futures to cause-effect chains overdetermining existence in education and social life more broadly. *Indigenous Futures and Learnings Taking Place* disrupts the common sense of "futures" in education or "knowledge for the future" by examining the multiplicity of possible destinies in coexistent experiences of living and learning.

Taking place is the intention this book has to embody and world multiplicity across the landscapes that sustain life. The book contends that Indigenous perspectives open spaces for new forms of sociality and relationships with knowledge, time, and landscapes. Through Goanna walking and caring for Country; conjuring encounters between forests, humans, and the more-than-human; dreams, dream literacies, and planes of existence; the spirit realm taking place; ancestral *luchas*; Musquem hən̓q̓əmin̓əm̓ Land pedagogies; and resoluteness and gratitude for atunhetsla/the spirit within, the chapters in the collection become politicocultural and (hi)storical statements challenging the singular order of *the* future towards multiple encounters of all that is to come. In doing so, *Indigenous Futures and Learnings Taking Place* offers various points of departure to (hi)story educational futures more responsive to the multiplicities of lives in what has not yet become. The contributors in this volume are Indigenous women, women of Indigenous backgrounds, Black, Red, and Brown women, and women whose scholarship is committed to Indigenous matters across spaces and times. Their work in the chapters often defies prescriptions of academic conventions, and at times occupies them to enunciate ontologies of the not yet. As people historically fabricated "women," their scholarly production critically intervenes on time to break teleological education that births patriarchal-ized and master-ized forms of living. What emerges are presences that undiscipline education and educationalized social life breaking futures out of time.

This book will be of great interest to students and scholars of Indigenous studies, future studies, post-colonial studies in education, settler colonialism and coloniality, diversity and multiculturalism in education, and international comparative education.

Ligia (Licho) López López is a Senior Lecturer at the Melbourne Graduate School of Education-University of Melbourne, Australia.

Gioconda Coello is an interdisciplinary scholar and Doctoral Candidate in Curriculum and Instruction at the University of Wisconsin-Madison.

Routledge Research in Anticipation and Futures

Series editors: Johan Siebers and Keri Facer

The Promise of Nostalgia
Reminiscence, Longing and Hope in Contemporary American Culture
Nicola Sayers

Indigenous Futures and Learnings Taking Place
Edited by Ligia (Licho) López López and Gioconda Coello

Sustainable and Democratic Education
Opening Spaces for Complexity, Subjectivity and the Future
Sarah Chave

https://www.routledge.com/Routledge-Research-in-Anticipation-and-Futures/book-series/RRAF

Indigenous Futures and Learnings Taking Place

Edited by
Ligia (Licho) López López and
Gioconda Coello

Routledge
Taylor & Francis Group
LONDON AND NEW YORK

First published 2021
by Routledge
2 Park Square, Milton Park, Abingdon, Oxon OX14 4RN

and by Routledge
52 Vanderbilt Avenue, New York, NY 10017

Routledge is an imprint of the Taylor & Francis Group, an informa business

British Library Cataloguing in Publication Data
A catalogue record for this book is available from the British Library

Library of Congress Cataloging-in-Publication Data
Names: López López, Ligia (Licho), editor. | Coello, Gioconda, editor.
Title: Indigenous futures and learnings taking place / edited by Ligia (Licho) López López and Gioconda Coello.
Description: New York : Routledge, 2021. | Series: Routledge research in anticipation and future studies | Includes
bibliographical references and index. |
Identifiers: LCCN 2020032290 (print) | LCCN 2020032291 (ebook) |
ISBN 9780367894603 (hardback) | ISBN 9781003019299 (ebook)
Subjects: LCSH: Human ecology–Philosophy. | Indigenous peoples. |
Manners and customs. | Forecasting–Study and teaching.
Classification: LCC GF21 .I5135 2021 (print) | LCC GF21 (ebook) |
DDC 304.201–dc23
LC record available at https://lccn.loc.gov/2020032290
LC ebook record available at https://lccn.loc.gov/2020032291

ISBN: 978-0-367-89460-3 (hbk)
ISBN: 978-1-003-01929-9 (ebk)

Typeset in Bembo
by Taylor & Francis Books

To the sacha, the story tellers, and our grandmothers who have been weaving spiraling futures

Ligia and Gioconda

Contents

Figures

Contributors

Ligia (Licho) López López is a Senior Lecturer at the Melbourne Graduate School of Education-University of Melbourne, Australia. She is the author of *The Making of Indigeneity, Curriculum History, and the Limits of Diversity* (Routledge, 2018). Her work has appeared in Race Ethnicity and Education, British Journal of Sociology of Education, and Curriculum Inquiry. Her upcoming books include *Migrating Americas* and *Growing up Antiblack in Latin America and the Caribbean.*

Gioconda Coello is a doctoral candidate in the department of Curriculum and Instruction at the University of Wisconsin-Madison. Her research is interdisciplinary and looks at the history of ideas in education and their relation to the politics of being, Indigenous, Brown and Black lives, and environmental education in Latin America and Southeast Asia. Her work has appeared in Revista Asia-America Latina.

Jo Anne Rey is a Dharug community member, currently delivering Dharug custodial-centred undergraduate curriculum: *Dharug Country: Presences, places, and people*, for the Department of Indigenous Studies, Macquarie University, North Ryde, NSW, Australia. Her 2019 doctoral thesis, *Country Tracking Voices: Dharug women's perspectives on presences, places, and practices*, was undertaken within the Department of Educational Studies, at Macquarie University. She is preparing post-doctoral research centred on activating knowledges across 3 sites on Dharug Ngurra (Country).

María Jacinta Xón Riquiac, Maya-K'iche' from Guatemala. Founder and researcher of the TUX Project: Gourmet Kitchen of Origins, in Chichicastenango, El Quiché. This project is dedicated to investigate the science of Indigenous women, the reproduction of life from in relation with the land, nutritional logics, chemistry in pre-Hispanic and pre-industrial cooking. The project attempts to re-signify the domestic space as one of oppression/resistance. She is an anthropologist and historian of science.

Nicole Tu-Maung is an interdisciplinary environmental scholar working in education, sustainability, and communications. She has held a faculty position at the Parami Institute in Yangon, Myanmar and received her MS in Environment and Resources from the University of Wisconsin-Madison.

Grace Pimentel Simbulan is a documentary filmmaker whose works have been screened throughout the Philippines, South Korea, Japan, and China. Her first feature film appeared on CNN Philippines Top 10 Films of 2019. She is a 2nd year MA student at the Center for Southeast Asian Studies-University of Wisconsin-Madison where her research focuses on the various forces affecting Indigenous groups in the Southern Philippines.

Ñusta Carranza Ko is an Assistant Professor of Global Affairs and Human Security in the School of Public and International Affairs at the University of Baltimore. Her research interests include cross-regional research on human rights and transitional justice processes in Latin America and East Asia, including policies of memorialization in Peru and South Korea and questions of Indigenous peoples' rights and Indigenous identities in Peru.

Jan Hare is an Anishinaabe scholar from the M'Chigeeng First Nation, located in northern Ontario, Canada. Her research is concerned with centering Indigenous perspectives within K to 12 and higher education. She is Professor and Associate Dean for Indigenous Education in the Faculty of Education at the University of British Columbia, where she also holds a Canada Research Chair in Indigenous Education and Resurgence. She also leads the Indigenous Teacher Education Program – NITEP.

Christine Bridge is a settler-scholar in the Department of Language and Literacy Education at the University of British Columbia (UBC). She completed her PhD in Literacy Education at UBC, and her research interests include place and land-based pedagogies, decolonization in teacher education, and literacy across the curriculum. Christine has taught at all levels in K-12 schools and in higher education.

Amber J. Shilling is Anishinaabekwe from Mnjikaning First Nation on her father's side and Polish-English on her mother's. She completed her PhD in Educational Studies at the University of British Columbia (UBC) and her interest areas include: Indigenous youth advocacy; technology-based Indigenous language, culture, and identity engagement; and decolonizing education.

Roberta Hill (aka **Roberta Hill Whiteman**) (1947-), Oneida, is a poet, fiction writer and scholar. Poetry collections include *Star Quilt* (Holy Cow! Press, 2001); *Philadelphia Flowers* (Holy Cow! Press, 1996) and *Cicadas: New and Selected Poetry* (Holy Cow! Press, 2013). Her poetry has appeared in magazines and anthologies, including *Sing: Poetry from the Indigenous Americas* (2011). Her most recent short story, "Reading the Streets" appeared in Narrative Witness #2, "Indigenous Peoples: Australia-United States" online: https://iwpcollections.squarespace.com/nw2-roberta-hill.

Acknowledgements

We want to start by acknowledging the Countries that sustain us in all possible ways. Just like ourselves, this book is possible thanks to the multiple relations with people and more-than-people beings that have over the years taught us, moved us, shaken us, and sensitized us to the tensions of being part and existing in the colonized places of this planet. The native trees and grasses, behind which the cold Narrm sunsets hide, have been our sanity. In the midst of the vertigo that have been these times, Country has been the stable ground on which this book could gather and rise. We salute our ancestors who continue to watch over us. We are eternally thankful for their continuous company in this world by day and other worlds by night.

The depth and shape of the book became possible thanks to the labor, thought, dedication, dreams, and understandings the authors brought to these pages. Our sincere gratitude to the unstoppable women-scholars who despite the pandemic, personal illness, and global unrest came through with their work. We feel incredibly privileged and humbled to have been in your company, through your thoughts, in conversation, during one of the most difficult, though not new, times of our lives. Your ideas have enriched our days and provided us with a space for hope amidst the trying times in which the book finally took place as a finished collective work. Thank you for your communal worlding of futures at a time most needed.

The reviewers and readers who are also part of this collective endeavor deserve special thanks. Under intensified pressure and with multiple obligations, they gifted their time to care for and nourish this project. They offered each chapter valuable insights that strengthened the work and commitment they have toward new worlds. Maltyox Edgar Esquit and Marta Juana López. Maarubaa nginda Nikki Moodie. Migwéch Kyle Whyte. Myfel Paluga Andrea Ragragio, maraming salamat. Ahéhee' Valerie Shirley and Aresta Tsosie-Paddock. Mil gracias Carolina Cabezas, Sarita Galvez, Myriam Espinosa-Dulanto, and Elena McGrath. Jane Ferguson and Thomas Patton, Coll Thrush, and Theodore Karamanski, thank you. For all your language editorial and translation work, and importantly for your patience, thank you Oscar Tantoco Sequiña, Myron Medina, Gabriela Gaus, and Jorge Bayona. Many thanks to David Korfhagen for producing the index within a relatively short time frame.

Special thanks to Matthew Galway for your comments, readings, continuous language advice and copyediting of several communications and writings, and for the emotional support.

Our trabajo is driven by our ayllu (family and family-like friends) and all the people who are part of our paths and curiosities.

Gracias Fausto Andrés Coello, Gioconda Baquero, Fausto Coello, Milena Coello, Benjamin Coello, Jessica Coello, Ron Ringlund, Jerry Ringlund and Bonnie Ringlund, for being such an important part of what motivates and sustains my happiness, health, studies, and research. Yupaychani yachachik Armando Muyolema. Ñukanchik rimaykunaka wiñaypak ñuka yuyaywan kumpañankami. Warmikuna, karikuna, ñañakuna, turikunapish kankunapa rimaykunapish yachaykunapish ñukawan karashkamantami yapa yupaychani, nini. Pagarachu sachalla! Kanmanta ñuka samayukmi kani. A mi pueblo la Majagua, al Rio Sinú que nos dejó su agua en tantos viajes por la Majagua, al Caribe, lugar de multiples relatos de morenidad y negritud resplandeciente, y a todes ustedes querida familia Maria (Ciro) López, Francisco López, Leandro López, y Leonardo López gracias por seguir siendo mi foco, con o sin luz eléctrica. Y a todas nuestras familias extendidas, porque algún día esperamos que lleguen a estas páginas, gracias por ser las primeras masas sobre las cuales se apoya toda la labor que hemos podido dar a este y otros trabajos.

My Heart Listens: A Foreword

Inés Hernández-Ávila

"As in the time of Netzahualcóyotl, this is no bed of roses."– Nancy Morejón[1]

I am writing from Patwin land, where I live and work. I begin this foreword with a distinguished Black Cuban woman's voice, to situate the starting point of the almost 530-year story of these Americas, known by many other names in the languages of the pueblos originarios of this hemisphere. Morejón's poem represents a weaving of moments between the then and now, tying her own country's history to central Mexico, Tenochtitlan, where Cortés arrived after Cuba. The "then" in Morejón's poem is, of course, a recent moment compared to the longevity of Indigenous peoples on their lands in this hemisphere and globally. But the "then" she references is like our "now", in that some humans more than others are experiencing widespread genocidal devastation, this time due to the pandemic, and everyone is witness and/or participant to intense movements for social justice, and earth justice, as in the time of original invasion and dispossession, when there were those who exercised their agency to insure their people's survivance, as Gerald Vizenor would say. Implicit in Morejón's line is the fact that we are still here, those for whom history has indeed not been a "bed of roses." We are here, rooted in and ever remembering/intuiting/awakening to our original teachings, and our awareness over the centuries is widening and becoming more inclusive every second. And more and more, as with the writers in this collection, we are sharing our stories, dreaming futures by remembering, through the earth's "vibrational memory" (Bilawarra Lee, quoted in Jo Anne Rey).

The pandemic and quarantine have created a space for reflection—an enforced "slowing-down", an inadvertent quiet time, some might say a sacred time, for those who have allowed themselves this peering into the mirror of their lives in relation to all that is. *Indigenous Perspectives on Futures and Learnings: Taking Place* speaks eloquently to the issues facing humanity in ways that will, perhaps, surprise some, and resonate with those of us who are (intimately) familiar with the messages these writers offer. The editors note that "the impetus of this volume requires the personal." This foreword is quite personal for me, because I hear these women's voices, I feel them, and upon reading these essays, I found myself reaching a place of (re)conocimiento as I contemplated the words of each writer.

The collection's dedication says: *"To the sacha, the story tellers, and our grandmothers / who have been weaving spiraling futures."* From this point I was pulled in. Spiraling. I went into my studio to paint recently. Often I paint not knowing what will appear. I see what emerges as I'm painting. This is calming for me because I am centered and the world drops off. I enter the space of creation. What appeared on the canvas was an impressionistic image of a woman in swirling, spiraling dance, her body turning, curving, her arms outstretched as if to reach with all her might to effect something, to will into being a healing, a transformation. I was stunned at how she emerged from me. I saw what I have been feeling. I have been swirling, spiraling, flowing upon eddies of grief, getting drawn into whirlpools of sorrow, the blur of movement bringing me back and forth in history, with history, back and forth from grief to longed-for jubilance. The spiraling of time and history. "The coexistence of place, time, and space," as Ñusta Carranza tells us. "Anticipatory consciousness," as Rivera Cusicanqui says. This collection offers readers discrete places, emerging from the cultural authority (Rey) of each of the writers, to see what I trust many readers have been feeling.

I have often thought of collections (of essays, of poetry) as creating, in the pages, a community of writers with each other, and with the readers. The editors call this collection "a territory of stories," which I appreciate. The writers are enacting a place of convocation, through an embodied (re)claiming of territory in each word, on each page, and in each chapter—the (re)claiming is global, and we are introduced to enlightened teachers, writing for healing, creation, liberation, writing against invisibility and Western "normalizing," and even writing beyond death and grief. The thread of double consciousness weaves throughout. "Goanna Walking"—walking with steps on the left and steps on the right (Rey). "Ishkay yachay, two knowledges" (Coello). "Land education" and contesting settler narratives with Indigenous knowledges, in physical places" (Hare, Bridge, and Shilling).[2] Learning that subverts yet still addresses official, institutional boundaries (all). As I consider the voices in this collection, I think they might agree with Mapuche poet and scholar, Elicura Chihuailaf Nahuelpán, when in his *Message to Chileans*, he addresses the nation, saying, "I think there has been an unwillingness to begin a Conversation about how to systematize and analyze the culture of the Other, the one who is different. Are we the Other, or are you?"[3]

For Indigenous peoples, Black peoples, other peoples of colour and the LGBTQUIA communities, "normal" is an oppressive, repressive, dehumanizing "othered" state—systemic racism, heteropatriarchy, assaults on the earth—even in the midst of multiple, but occasional personal and collective triumphs. Voracious global capital, or what my colleague, Liza Grandia calls "corporate colonialism"[4] has determined that most humans are trained to be workers to serve, to abide by the clocks, calendars, requirements, rules, and schedules that are imposed to confine, contain, restrict, and control us, and to enforce violations of the earth, while others are raised to be privileged, to have license to do as they will, with other humans, with the more-than-humans, with the earth, the waters, and the skies themselves. What the rapacious ones ignore is the fact that the earth and the universe have been and are witness to everything. We are being shown that it is time for change. This collection is about claiming place, futures taking place now, rather than "knowing our place," or "staying in our place," according to old imperial, colonial, corporatized scripts. It is about agency, responsibility, consciousness, and the sovereignty and autonomy of all of life. Jo Anne Rey asks the vital question, "What kind of presence do I want to embody as an actor in the collective future?" Each of the writers, Gioconda Coello, Ligia López López, Jo Anne Rey, María Jacinta Xón, Nicole Thuzar Tu-Maung, Grace Simbulan, Ñusta Carranza Ko, Jan Hare, Christine Bridge, Amber Shillings, Roberta Hill, offers a path, enacting the ways in which we can make a difference.

Dreaming emerges in the chapters by Jo Anne Rey, María Jacinta Xón Riquiare, Grace Simbulan and Gioconda Coello, which corresponds to Native American and Indigenous Studies scholars, such as Jo-Ann Archibald, who are writing about dreaming as a methodology, as well as storying.[5] The theme of the "good life" is one that is familiar and welcome, and Gioconda Coello's essay on Kichwa stories I know will resonate with the similar but discrete ideas, such as the concepts "bimaadiziwin . . . the good life, . . . a long and healthy life" of the Anishinaabe (Lawrence Gross); the "good life" of the Sto:lo (Lee Maracle); and lekil kuxlejal, "to live in a dignified manner," of the Maya Tsotsil of Chiapas (Mariana Mora).[6] Each of the chapters are learning moments full of possibility and visioning that are being activated and that, for readers, can contribute to their own enacting, regarding the earth's healing powers, (re) newed understanding of time (Xón Riquiac), the significance of each of our hi/stories when we comprehend that our lives do not take place in a vacuum—each of us, like Ñusta Carranza's father, in his "cycles of journeys" is a bearer of history, if we would only step back to see how. Jo Anne Rey writes of "walking towards sustainable and ethical futures" and Gioconda Coello tells us about Belgica's teachings, and how we "need to learn how to walk in the forest," gently, tenderly, listening, always listening.

The concept of "co-becomings with more-than-human beings" is dear to me. Early in this pandemic, I wrote, "la tierra respira cierta tranquilidad/los seres sensibles nos miran/a ver si sabemos co-vivir"—"the earth breathes a certain tranquility/the sentient beings observe us/to see if we know how to truly live together." I love the animals coming out to stroll along city streets, to explore, feeling their jubilance, their curiosity, their own anticipatory consciousness. They have been forcibly relocated, contained in their own way, being (increasingly) denied their territories, being denied the mappings that they know, that are part of their nature, their internal geography, having to always watch, always be on the alert for predatory behaviour, for corporate incursion, for disruption. As the smog lifted from major cities, as waters cleared up, and traffic was considerably altered, earth, water, sky relations took immediate notice, sensing change. In my own back yard, since the shut-down, my husband and I suddenly became aware of a community of stray cats, three with collars, one feral, who came to take residence in our yard. We began putting out food and water for them. They joined the society of squirrels in our trees, and a family of possums who live outside. They are truly teachers and sacred helpers, as Jo Anne Rey, Nicole Tu-Maung, and Roberta Hill have indicated. They share, living in community, recognizing each other's needs. Each cat, possum and sometimes a bluejay or two, eat just what they need and leave the rest for the others (the squirrels have acorns galore). They even seem to keep a schedule—who eats when, who gets what time to snooze on the deck in back, or in the front yard, or in the coolness of a corner. There was one cat that tried to dominate at first, but peer pressure made her back off, and she is actually much calmer now, because she knows she has her own place in this community. Each one comes and goes, making place for themselves in and around our yard. We live in harmony with each other. I felt joy when I read the found poetry in Rey's declaration from her own cultural place: *"And across the site, it rained, and the kangaroos returned through the mists in the distance. This was our Dharug ecological-sovereignty being enacted."* I feel that the more-than-human beings are coming out to communicate with us and show us the way, and this collection helps us to see this.

I appreciate the organization of this collection, and the placement of Roberta Hill's chapter on the Haudenosaunee Thanksgiving Address as the last chapter. *Atunhetsla.* Life and spirit, a cosmovision that embraces the universe and affirms the kinship we have with all our relations. I am happy to say that the beloved Jake Swamp came to UC Davis in 1984 to plant a Tree of Peace that stands next to our administration building. The Native community on our campus understands the significance of this act and we are proud to be connected in this way to the Haudenosaunee. As Hill notes, via Jonathan Patz, habitat destruction and climate change are driving COVID-19, although we know that before COVID-19, we were well on our way to heartbreak from the damage we have done to this planet. Destruction has consequences. But if we remember to always, always give thanks for life, for our growing awareness, for the lessons we have learned and the blessings we have received, from the tiniest to the grandest, in our thanksgivings we are recognizing that we are each part of the whole of the universe—and in doing so, accepting our responsibility for our futures taking place. As Morejón wrote, we are living in a time that is not a bed of roses, but even if it were, roses do have thorns, just as their presence and their fragrance are so full of beauty. Love-and-pain. Grief-and-beauty. Death-in-life and life-in-death. Grace Simbulan shares an excerpt of the powerful poem by Bo-i, "We Who Dare": "We who dare must lick our wounds/Turn to Magbabaya to lift our thorns/We may have tilt on the losing side/But on his grace we can depend/We who dare are his seeds on earth/We who dare must keep moving on." I am so content to close with a poem, after opening with one. And as a final note, in honour of all of the contributors to this volume, I agree, we *must* be the ones who dare.

Notes

1 Nancy Morejón, "A dream of reason produces monsters," *Bomb*, Issue #14, 1 January 1986. Translation from Spanish by Zoë Anglesey.

2 At UC Davis, Native American Studies is doing some of this work, ensuring that the campus is known and remembered as Patwin land, not merely through a land acknowledgement statement, but through physical places that we have worked to create, such as a Contemplative Garden that honours the Patwin people of today, and their history; a marker with the words of Patwin elder, Bill Wright etched in stone; and the naming of local plants with their Patwin names, in our Arboretum. We are not as advanced as those working at UBC, but we are on this path.

3 Elicura Chihuailaf Nahuelpán, *Message to Canadians* (Victoria, B.C.: Trafford Publishing, 2009), p. 79.

4 Liza Grandia, my colleague in Native American Studies, teaches a course titled "Corporate Colonialism," which focuses on the impact of the corporation and neoliberalism, military and intelligence agencies, and the price of progress and modernity, and more on Indigenous peoples and all peoples.

5 Jo-Ann Archibald discusses dreaming in *Storywork: Educating the Mind, Heart, Body, and Spirit* (UBC Press, 2008).

6 Lawrence Gross, *Anishinaabe Ways of Knowing and Being* (Burlington, VT: Ashgate Publishing, 2016), pp. 205-206; Lee Maracle, *Memory Serves: Oratories*, ed. Smaro Kamboureli (Edmonton, A.B., NeWest Press, 2015), pp. 220-229; Mariana Mora, *Kuxlejal Politics: Indigenous Autonomy, Race, and Decolonizing Research in Zapatista Communities* (Austin: University of Texas Press, 2017), pp. 18–23.

Futures taking place

Gioconda Coello and Ligia (Licho) López López

We, Gioconda Coello and Ligia López, would like to acknowledge that these pages are written on Native American and Aboriginal lands that have never been ceded—the lands under Ho Chunk and Wurundjeri and Boon Wurrung Peoples respectively. We acknowledge the Indigenous and Black country that makes our life and scholarship possible where we remain uninvited guests from other stolen lands; Ligia at the shores of the Caribbean Sea and Gioconda on the Highlands of the Andes. In the collection that this book is, of stories, histories, futures formation *and place*, this acknowledgement of country is an imperative to situatedness—of us in place, as the futures that we already are. For our ancestors—who for the past 528 years have been called "indians" but who, for far longer than Chronos' time can tell, are just one with nature and the universe—we are the future in this present. And yet it is this conventional order of time that we cut through and across to invent other rhythms of time. So,

well

come,

welcome to these pages. They comprise a territory of stories woven by women from different places across the globe and across fields of knowledge. This territory is a space to gather together to examine the multiple journeys, times and knowledges that take place in the world from Indigenous peoples' and more-than-human perspectives. The contributing authors in the volume interrogate the colonial conditions that overdetermine futures and learnings, and narrate ways of existing, resisting, struggling, and caring for peoples, spirits, and lands. Through those narrations the authors in this book challenge, as Vanessa Andreotti[1] has put it, the torpor stemming from an education broadly understood, which limits people to the knowing and being they have been socialized into. We bring stories to sensitize people to other (im)possibilities.

The editors are us, two women from the place hideously-named after a colonizer—the Americas. Ligia is a descendant of dwellers of the Sinú Riverbanks in the Caribbean coast (of the place shamefully named) Colombia.

Gioconda is a descendant of peoples of the Andean highlands (of a place named after European imaginary world lines, Ecuador). The contributors are Jo Anne Rey, a Dharug woman in Dharug Country (Sydney, Australia); María Jacinta Xón, Maya K'iche' in Chichicastenango (Guatemala); Nicole Thuzar Tu-Maung, descendent of peoples of a place called Burma by the British and Myanmar by a military junta and a first-generation child on Haudenosaunee lands (New York, U.S.A.); Grace Simbulan, descendant of the peoples of Mariquina (Philippines); Ñusta Carranza Ko, descendant of peoples of the Andean highlands in Quitaracsa (Peru) and generations of families from the Koreas; Jan Hare, Anishinaabe woman from the M'Chigeeng First Nation (Ontario, Canada); Christine Bridge, descendent of settlers on Okanagan lands and residing on Musqueam territory (British Columbia, Canada); Amber Shillings, Anishinaabe woman from Mnjikaning First Nation (Ontario, Canada); and Roberta Hill, Oneida woman on Ho-Chunk and Oneida territory (Wisconsin, U.S.A.).

This book came about from the thirst of hearing and reading experiences and understandings of futures that escape chronological time[2] and the urge to plan existences. Time orders education in a fundamental way not only because of school calendars, class and testing schedules and the management of advancement through curriculum milestones, but importantly because education is meant as a way to change the person and equip them for times to come. This is clear in public education systems, which are linked to the production of future citizens and societies. That kind of time regime plans or anticipates people into entanglements of values to face the uncertainties and the fears-hopes of societies. Anticipatory regimes abide in the ontology of the not-yet.[3] Thus, anticipating implies an orientation of the self to inhabit time out of place by knowing that which has not yet become.[4] That is, anticipation implies "a performative process of rendering the future actionable."[5] In the case of schooling, students become an actionable domain as future citizens/actors to (re)produce society and ensure security to the kind of life that is valued. For instance, students (re)produce societies that make other of Indigenous and Black people and 'ancestralize' their relation to their territory when they learn and act in the world following curricula that normalize certain languages, cultural values, and homogenous national identities. That actionable domain is 'activated' on the basis of data, predictive curricula, testing, and educational policy to intervene in the learning experiences and lives of people.[6] As Sarah Amsler and Keri Fracer have proposed, the educational or schooling subject within this regime becomes a "target of multi-governmental-level algorithmic decision-making"[7] and the instrument for avoiding risks, and 'enhancing' the possibilities of the self for "human capital."[8] However, anticipation does not need to be determined by reducing risks to the moral expectations of some and "the instrumentalizing of life for another's power and profit."[9] Anticipation can be a critical engagement with time to break with teleological thinking by opening the relationships and perspectives of time in many of its instances and disrupting linear progressions between learning and becoming. As Silvia Rivera

Cusicanqui[10] phrases it, an "anticipatory consciousness" or "principle of hope" can emerge from the present informed by the spiral movement of futures and pasts. The continuous feedback in those spirals makes possible to explore and discern what conforms time and to expand the hopes for otherwise futures. This is what this book has to offer by approaching what has not yet become from Indigenous perspectives.

We acknowledge that there is not one single trajectory for what becomes and there is never a single version of (hi)story. In this sense, the book is occupied with the multiplicity of future(s), according to the diverse and coexistent experiences of living in the world. Future(s) emerge through the chapters as the continuation of presences; generational journeys creating and breaking cycles; co-becomings with more-than-human beings; rebirth; new beginnings; pathways to liberate people from cycles of suffering; an undulation of the time serpent; and stories and dreams of what is yet to become. They are stories of survivance understood as "an active sense of presence over absence [...] the continuation of stories, not a mere reaction, however pertinent. [...Stories of] renunciations of dominance, detractions, obtrusions, the unbearable sentiments of tragedy, and the legacy of victimry."[11] Future(s) in this (con)text thence speak of the intention of continuance and the narration of presences in wavy stories and visions spoken in agential intersections.

Similarly, learning is proposed in this book as a capacious concept to embrace the multiplicity of relationships that bring about knowledge. As Paul Hager[12] suggests, learning as belonging to schooling is a more familiar and often privileged understanding of what learning can mean but by far not the only one. Often education systems depend on the "reasonable" assumption that knowledge and skills can be acquired through a well-defined and determined path. However, learning is a social, cultural and political experience that flows through and beyond schools. In this book we approach education as a way of survival understood in relation to what sustains people's lives,[13] and survivance as the active sense of presences. Learnings belong to the will to living in a respectful entanglement of the self, other beings, and the worlds that sustain life.

Learnings are conceptualized through the chapters as transformative experiences and actions with the purpose of caring. As Jo Anne Rey (Chapter 1) explains, learnings are embedded through place, allowing room for cultural and transgenerational memory as well as intuitions to enable reflexivity and consciously become together with the land and all other agencies. Nicole Thuzar Tu-Muang (Chapter 4) and Grace Simbulan (Chapter 5) remind us that those agencies include spirits and ancestors who communicate through dreams for the creation of communal discourses and negotiation of political power by asserting or resisting knowledge, practices, and ways of life. Engaging with Indigenous knowing, as Jan Hare, Christine Bridge, and Amber Shilling (Chapter 7) assert, also requires an engagement with processes of reconciliation and decolonization. Experiences, territory and teachers have the potential to denaturalize the colonial arrangements which have either denied or commodified Indigenous knowledge (see Chapter 6 by Ñusta Carranza). Learnings thus are pedagogical encounters that can continue to repeat collective dispositions and existential imperatives that draw violent boundaries or can change understandings and draw paths towards respectful good living (see Chapter 2 by Gioconda Coello and Chapter 3 by María Jacinta Xón Riquiac). As Roberta Hill (Chapter 8) rightly suggests, learnings come with images and stories of death, but also with power to heal in the stories of people and their relationships with Earth.

This book looks at how future(s) and learning(s) inhabit social, spiritual and physical landscapes. Landscapes are the common place that comes about in ideas, stories, narratives about the nature of reality and the self, as well as the climate and geographies that make those stories and ideas possible.[14] The experience of that common place is primordial to what we understand as learnings, as it provides people with conceptualizations of knowledge and reality while simultaneously provoking shifts in those conceptualizations.[15] *Taking place* implies a recognition of how future(s) and learning(s) come about and move across the landscapes that sustain the life of people and the life of life more broadly: Their personal (hi)stories, environments, memories, dreams, desires, spiritualities, communities and societies. *Taking place* is the intention this book has to open a space for new socializations and relationships with knowledge in those landscapes.

Indigenous perspectives, we propose, are able to open those spaces. We understand Indigenous to be a genre[16] of existence that has been differentiated through politicocultural and (hi)storical statements. Those statements are relational to the stories of empires and kingdoms which overdetermined their own existence as the norm and in doing so justified the use of their power to displace people, narrate them as backward, and change their destinies. These narratives of normalcy and advancement have continued and are present in the twenty-first century, linked to the production of ideas of progress and development privileging the thought and practices stemming from the sciences and national governments.[17] In relation to a tacit national *we*, these narratives produce the Indigenous/Native/Aboriginal/Black questions that turn people into research (s)objects to control, "help," and even admire *them*. [18] Even though naming the Indigenous existence carries a tacit normalized existence, the authors and editors of this book do not see Indigeneity limited to that relation. The stories presented here make self-statements and statements through relationships centered on listening, based on respect, and opening space for the stories that often feel constrained within the heritage of modernity as described by Bruno Latour.[19]

In that spirit, expanding the meaning of futures and learnings from Indigenous perspectives requires altering the conventional modes of presentation of what constitutes a book and its chapters. We recognize that the recent invention of objectivity (which among other things requests a bracketing out of the personal) maintains a stronghold on how meaning is to be negotiated in intellectual spaces. The impetus of this volume requires the personal—the meanings and stories of people and collectives—to take centre stage. We know these modalities are capacious for the reimagination of educational relations with what is yet to become. From that space, the authors consider: How are expectations, ideas, feelings, dreams of future(s) reached through Indigenous people's ways of knowing and practices? How do they translate into learnings of the world, communities and (hi)stories that are vital to continue to be? How do those learnings and futures from Indigenous perspectives take place and how are they embodied around the world? The answers come through experiences, dreamings, poems, conversations and academic curiosity.

Chapter one brings the presences of Dharug Country (Sydney, Australia) to demand attention over the pressures undermining ecological systems and processes that compromise sustainable futures. In weaving ancestral values, stories and connections with/in/as contemporary presences, places and practices, seven Dharug women's yarnings produce a Dharug Ngurra web of belonging, caring and connecting. "Goanna walking" opens a third place that recognizes the intercultural nature of existence, as embedded diversity within locales. Further, Goanna walking brings other-than-humans directly into pedagogical domains as an opportunity for diversifying agency. Jo Rey proposes in this chapter that engaging Goanna spreads transformative trailing tail-tales, fostering future co-becomings and learning(s) that can be shared locally, nationally and internationally for the benefit of all sentient beings.

Chapters two and three question the narration of time and futures. In Chapter two, Gioconda Coello contrasts the logics underpinning the concept of "good living" as espoused by the Ecuadorian government with the logics present in Kichwa stories about future and "good living." She argues that the Kichwa ways of learning about future(s) to live well disrupt extant bureaucratized conceptions of "good living." She draws on Sylvia Wynter's ideas of the *homo narrata* [20] to propose stories to be the narration of truths from diverse ontological possibilities according to the ways of life and thinking of the (human and more-than-human) story tellers. This chapter disrupts the narrative about "good living" produced in relation to the governmental planning of "good life" and open conversations about a "good living" for the future(s) embedded in Kichwa stories about the world, the land, dreams and everyday living. María Jacinta Xón Riquiac, in Chapter three, asks how is time conceived, lived and co-lived in the quotidianity of millions of Indigenous peoples in Mesoamerica? How to explain chu wuch q'ij saq/the existence of time beyond the Mayanist discourse? Given that chi ma je'/like that is because it is like that, is foundational to Mayan quotidianity in the present, how to explain the existence of time through the exercise of a narrative that says chi ma je'? She engages with these questions through the story of a girl who dreams of the future and other k'iche' experiences of the chi ma je'/like that it is because it is like that.

Indigenous spirits bring learnings in Chapter four, in which Nicole Thuzar Tu-Maung analyzes Snake Temples, Theravada-Buddhist places of worship in Myanmar that are made notable through the veneration of live and captive Burmese pythons (Python *bivittatus*). The pythons who inhabit these temples are viewed as *Nats* spirits considered Indigenous to present-day Myanmar. The *Nat* spirits are specifically those associated with the *Naga*, a powerful serpent of the spirit realm. Through communications with these spirits, followers learn how to improve their current material/economic condition, deepen their spiritual practice and the preservation of the Buddhist religion. Tu-Maung argues that through these temples Indigenous *Nat* spirits and their believers produce future(s) for themselves within novel Buddhist imaginaries and sacred spaces. These future(s) are linked to a particular form of "prosperity Buddhism," one that is created by a community of people, spirits and snakes.

Chapters five and six draw on a single-person story to powerfully illustrate survivance and resistance of entire communities. In Chapter five, Grace Simbulan explores the contemporary role of dreams in the community life of the Obo Manobo, an Indigenous community in Mindanao, Southern Philippines. She follows the dreams of an Obo Manobo social and spiritual leader to argue that dreams are a crucial form of learning for resistance which bring to the present the knowledge of the ancestors to ensure continuance of the life of the Obo Manobo people-and-land. In Chapter six, Ñusta Carranza shares a reflection of the personal struggles of her father, a Quechua man from Quitaracsa, Peru. Her chapter shows us how education for "integration" into society has marked the generational journeys and making of communities' futures for Indigenous peoples. This chapter discusses Indigenous survivance based on the temporalities of learning and loss and the recurring nature of the struggle against discriminatory practices within education which transcend linear conceptions of time. The memories of the author's father weave a cyclical spiral time frame that connects the past, with the present and future stories of his *ayllu* (family) and relationally of other Andean communities.

In Chapter seven Jan Hare, Christine Bridge and Amber Shilling highlight the pedagogical activity conducted with pre- and in-service educators through reflection on their storied relationships to history, land, time and place using the University of British Columbia (UBC) campus. Following their experiences of five sites, student teachers/learners reflected on the themes of Indigenous reclamation, resurgence and reconciliation. This pedagogical practice of learning with, on and from land and place seeks to guide learners in denaturalizing their understandings of colonial arrangements by empowering ancient and dynamic knowledge buried under post-secondary education developments. The chapter scaffolds for readers how they might adapt, modify and accommodate this pedagogical activity within their own contexts, giving consideration to specificities of time and place, making visible local Indigenous histories, knowledges and stories in curriculum and planning and attending to local relationships. Through the sharing of pedagogical practices that support learning from the elders and from the land and place, these chapters reaffirm a commitment to advance Indigenous priorities.

Engaging with the present times when "Mother Earth is sending us to our rooms" through a pandemic, in Chapter eight Roberta Hill reflects on the practices, prayers and thoughts in the Thanksgiving Address that give continuance to Haudenosaunee people and their futures through kinship. The author reminds us how the attitude and depth of colonization are pervasive in the present(s) and yet the spirit and philosophy of the Haudenosaunee Thanksgiving Address points to an otherwise way to relate to life, the planet, communities, and the world future(s). It is in the recognition of the interconnected relationships between the Earth, sky and all that is in-between, that a communal vision can help people let go of their resentments, generate gratitude and see the need for differences. The remembrance involved in the practices of the Thanksgiving Address focuses on life as a continual spiritual force that accompanies beings especially in facing death and hardship. Learnings from grief, memories, kin and the Earth strengthen responsible action and help people stay resolute in their desires for otherwise future(s) and their gratitude for *atunhetsla*/the spirit within.

Notes

1 Vanessa Andreotti, "(re)imagining education as an un-coercive re-arrangement of desires," *Other Education: The Journal of Educational Alternatives,* 5, 1 (2016): 79–88.
2 Ligia (Licho) López López, *The Making of Indigeneity, Curriculum History, and the Limits of Diversity* (New York: Routledge, 2017): 115.
3 Sarah Amsler and Keri Fracer, "Contesting anticipatory regimes in education: exploring alternative educational orientations to the future," *Futures*, 94 (2017): 6–14.
4 Vicanne Adams, Michelle Murphy and Adele Clarke, "Anticipation: Technoscience, life, affect, temporality," *Subjectivity*, 28 (2009): 246–265.
5 Ben Andersson, "Security and the future: Anticipating the event of terror," *Geoforum*, Vol.41, (2010): 229.
6 Ben Williamson. "Governing software: Networks, databases and algorithmic power in the digital governance of public education." *Learning, Media and Technology*, 40, 1 (2015): 83–105.
7 Amsler and Fracer, *Anticipatory Regimes*, 10.
8 Ibid, 10–11.
9 Sarah Amsler, "Gesturing towards radical futurity in education for alternative futures," *Sustainability Science*, Vol 14, (2019): 925.
10 Silvia Rivera Cusicanqui, "Ch'ixinakax utxiwa: A reflection on the practices and discourses of decolonization," *The South Atlantic Quarterly,* 111,1 (Winter 2012): 95–109.
11 Gerald Vizenor, *Survivance: Narratives of Native Presence* (Lincoln: University of Nebraska Press, 2009), 1.
12 Paul Hager, "Conceptions of learning and understanding learning at work," *Studies in Continuing Education*, 26, 1 (2004): 3–17.
13 Cynthia Chambers, "A topography for Canadian curriculum theory," *Canadian Journal of Education/Revue Canadienne De Léducation,* 24, 2 (1999).
14 Ibid.

15 Vanessa Andreotti, Amosa Faafoi and Margaret Giroux, "Shifting conceptualisations of knowledge and learning in the implementation of the New Zealand curriculum: Conceptual models and a preliminary analysis of data," *Waikato Journal of Education*, 15,1 (2010).

16 See Sylvia Wynter, "Unsettling the coloniality of being/power/truth/freedom: Towards the human, after man, its overrepresentation—an argument," *The New Centennial Review*, 3,3 (2003):257–337. Wynter proposes the genre of human as historical, cultural, economic and hereditary variations of the possibilities of being human narrated into being through descriptive statements of human. Wynter examines genres of human in relation to the overrepresentation of the biocentric ethnoclass genre of the human shaped by European history and the discourses of capitalism and White/non-White relation. Here we propose genre of existence to account for the more-than-human beings including spirits and land.

17 Ligia (Licho) López López, *The Making of Indigeneity, Curriculum History, and the Limits of Diversity* (New York: Routledge, 2017).

18 Ibid; Gioconda Coello, "Producciones narrativas, tesis culturales y ficciones: una historia del presente de Sumak Kawsay [Narrative productions, cultural theses and fictions: a history of the present of Sumak Kawsay]," in *Convergencias sobre la producción cultural ecuatoriana*, Manuel Medina and Norma González, eds (Quito: Universidad Técnica Particular de Loja, 2019).

19 Bruno Latour, *We Have Never Been Modern*, trans Catherine Porter (Cambridge: Harvard University Press, 1993). Latour conceptualizes the modern as a way of thinking which artificially divides nature and society and abides in tension between rationality and obscurantism and ideology and science. This kind of modern thought is grounded in the belief on the certainty of nature's laws, on humans' ability to make their destiny on their own and the objectivity of science versus biased Other ideologies.

20 Katherine McKittrick, ed. Sylvia Wynter, "Unparalleled catastrophe for our species? Or, to give humanness a different future: Conversations," *On Being Human as Praxis*. (Durham: Duke University Press, 2015): 9–89.

References

Adams, Vicanne, Murphy, Michelle, and Clarke, Adele. "Anticipation: Technoscience, life, affect, temporality." *Subjectivity*, 28: 246–265, 2009.

Amsler, Sarah. "Gesturing towards radical futurity in education for alternatives futures." *Sustainability Science*14: 925–930.

Amsler, Sarah and Fracer, Keri. "Contesting anticipatory regimes in education: exploring alternative educational orientations to the future." *Futures*, 94: 6–14, 2017.

Andersson, Ben. "Security and the future: Anticipating the event of terror." *Geoforum*41: 227–235, 2010.

Andreotti, Vanessa. "(re)imagining education as an un-coercive re-arrangement of desires," *Other Education: The Journal of Educational Alternatives*, 5 (1): 79–88, 2016.

Andreotti, Vanessa, Faafoi, Amosa, and Giroux, Margaret. "Shifting conceptualisations of knowledge and learning in the implementation of the New Zealand curriculum: Conceptual models and a preliminary analysis of data." *Waikato Journal of Education*, 15 (1): 29–48, 2010.

Coello, Gioconda. "Producciones narrativas, tesis culturales y ficciones: una historia del presente de Sumak Kawsay [Narrative productions, cultural theses and fictions: a history of the present of Sumak Kawsay]," in *Convergencias sobre la producción cultural ecuatoriana*. Edited by Manuel Medina and Norman González, 405–418. Quito: Universidad Técnica Particular de Loja, 2019.

Chambers, Cynthia. "A topography for Canadian Curriculum Theory." *Canadian Journal of Education/Revue Canadienne De Léducation*, 24 (2) (1999).

Hager, Paul. "Conceptions of learning and understanding learning at work." *Studies in Continuing Education*, 26 (1) (2004): 3–17.

Latour, Bruno. *We Have Never Been Modern*, trans Catherine Porter, Cambridge: Harvard University Press, 1993.

López López, Ligia (Licho). *The Making of Indigeneity, Curriculum History, and the Limits of Diversity*. New York: Routledge, 2017.

Rivera Cusicanqui, Silvia. "Ch'ixinakax utxiwa: A Reflection on the Practices and Discourses of Decolonization." *The South Atlantic Quarterly*, 111 (1): 95–109, 2012.

Vizenor, Gerald. *Survivance: Narratives of Native Presence*. Lincoln: University of Nebraska Press, 2009.

Williamson, Ben. "Governing software: Networks, databases and algorithmic power in the digital governance of public education." *Learning, Media and Technology*, 40 (1) (2015): 83–105.

Wynter, Sylvia. "Unparalleled catastrophe for our species? Or, to give humanness a different future: Conversations," In *On Being Human as Praxis*. Katherine McKittrick, ed. Durham: Duke University Press, 9–89, 2015.

Wynter, Sylvia. "Unsettling the coloniality of being/power/truth/freedom: Towards the human, after man, its overrepresentation – an argument." *The New Centennial Review*, 3 (3): 257–337, 2003.

1 Changing places

Weaving city learnings into Country futures

Jo Anne Rey[1]

Weaving ancestral values, stories and connections with/in/as contemporary presences, places, and practices, seven Dharug women's yarnings undertaken for the doctoral research project: *Country Tracking Voices: Dharug women's perspectives on presences, places and practices* (Rey, 2019), produced a Dharug Ngurra web of belonging, caring, and connecting. Through puppetry, art, dance, song, poetry and yarning, in places of ancestral significance, agentic Ngurra wove presences, places and practices into futures and learnings. In this chapter, Goanna walks us through new relationships with Ngurra-as-city within the academy that revive Dharug Aboriginal cultural ways of knowing, being and doing.

When Dharug Ngurra (Dharug Country) is the majority of cosmopolitan Sydney, Australia, pressures on ecological systems demand our attention as we witness places, presences and ecological processes being dangerously undermined. Recognizing reciprocity as an overriding moral human imperative to care for Country, so that sustainable futures foster continuing presences, is no longer only an Aboriginal custodial obligation. In the Anthropocene, and with devastating climate crises across the globe, all humanity is called to address this imperative.

Apprehending this need requires alternative ways of knowing, doing and being. Finding such alternatives requires relating ourselves in the Universe differently. Arabena positions this as "indigenizing" ourselves to the Universe, so that we can recognize our relational presence and dependency within the cosmos.[2] Such an approach opens pathways that decompose human-centric binaried narratives and their extinction industries. Using methodological knowledges gained from earlier research (known as "Goanna[3] walking") it is recognized and demonstrated how alternative ways of knowing, doing and being are continuing in Dharug Ngurra-as-city.[4]

"Goanna walking" is a third way approach that weaves between ancient sustainable values and those of existing human-centric paradigms and modernity[5]. It requires walking with steps on the left, those Indigenous ways of knowing, doing and being, wisdoms that have sustained Ngurra for millennia, and walking with steps on the right, involving modernity: the outcome of patriarchal, hierarchical and human-centric methods that continue to underpin globalizing systems that are driving extinctions today.

As such Goanna's trailing tail-tale opens a third place that recognizes the intercultural nature of existence, as embedded diversity within locales. Further, Goanna walking brings other-than-humans directly into pedagogical domains as an opportunity for diversity-agency. It opens conceptions of Country-as-city for dissolving binaried hierarchies and allows equitable justices across domains. Beyond this, it spreads transformative trailing tail-tales fostering future co-becomings and learning(s) that can be shared locally, nationally and internationally for the benefit of all sentient beings.

This chapter undertakes such a "Goanna walk" to demonstrate how walking towards sustainable and ethical futures can open possibilities for decolonized coexistences, even when Country is a city.[6] Such a decolonization process involves pedagogy as ethical ecological relationships. Three examples of place-based, experiential engagement illustrate the processes and knowledges that are being uncovered. They include site-based research at Shaw's Creek at the foot of the Blue Mountains, undergraduate site-based experiential-learning at Macquarie University, and the journey that resulted in the return of the Blacktown Native Institution, at Oakhurst to Dharug custodians in 2018. They involve past-presences, weaving reflexive in-situ webs of connection through shared times-tellings, and active public participations. Together this journey demonstrates Dharug revival, return of custodial Country and along the "Goanna walk" provides cultural purpose for perpetuity.

Opening Place

Perspective 1

All that has been has changed,
All that will be, will change,
It's how we manage the changes,
That makes the difference.
(Someone, sometime, somewhere)

Perspective 2

Always was, always will be, Aboriginal land.
(Many, sometimes, somewheres)

Perspective 3

Walking between: a third way

Dharug Ngurra (Country) as city: taking place

Warami Mittigar. Welcome into Dharug Ngurra. It is the place of belonging and the Country of the Dharug-speaking peoples. I acknowledge and pay my

Figure 1.1 Goanna Research Journey: Dharug Doctoral Thesis map burnt on Possum Skin. Jo Anne Rey, 2019

respects to Ancestors and Elders, past, present and emerging, and remember that they always have been, and always will be caring for Country.

Dharug have been, for many thousands of years, and continue to be, the first custodians of the lands, seas, sky, all the physical diversities, and all the metaphysical spiritualities within the area that covers the majority of the cosmopolitan metropolis of Sydney, NSW, Australia. Flying as a hawk around the boundaries of Ngurra, you would see below, the Pacific Ocean in the east, the northern shoreline of Botany Bay in the south, and the coastline heading north to the mouth of the Hawkesbury River at Broken Bay. Heading inland above the Hawkesbury River (previously known in Dharug language as the *Deerubbin*) you fly until reaching, in the north-west, the Colo River. Follow the Colo and then cross the land following the early afternoon sun, towards the western rim of the Blue Mountains and its east-west ridgeline. Today spread below are the townships of Medlow Bath and Blackheath. Turn toward the morning sun, following this ridgeline down to the Nepean River (still the Deerubbin in our language), and follow the river to where it turns east. Flying easterly overland you will reach the George's River, at which point you will turn and follow it back to Botany Bay on the coast. All within, and above, these mainly water boundaries is Dharug Ngurra. Caring for everything within this space is the traditional custodial responsibility of Dharug peoples and guiding the stewardship of this area is a continuing cultural practice, principle and imperative. Together, with the storying, the knowledges, the connections to places, the ceremony, Law and our Ancestors, Dharug cultural authority is maintained.

Following the associated dual impacts of colonial-settler invasion (post-1788) and the outbreak of smallpox (1789), Dharug community, culture, practices,

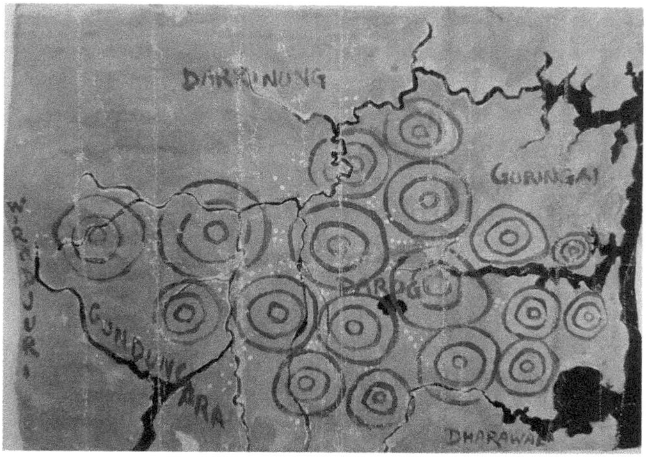

Figure 1.2 Dharug Ngurra Map.
Christopher Tobin, 2015

connections to food sources and language was decimated, but not wiped out. Dharug Ngurra today has a population of approximately five million human beings. Of these around 27,000 identify as Aboriginal and/or Torres Strait Island people.[7] Dharug custodians, who identify, represent a small percentage of these. Traditional custodial caring for Country in the city, therefore, is a major challenge.

In spite of the re-colonizing bureaucracies that are dominating many sectors of Aboriginal political space today, the Dharug community has been working diligently at getting on with the business of enacting Dharug cultural practices, yarning-up Dharug places of significance, conducting ceremony, educating our children, and continuing cultural authority that has been here for many thousands of years. Dharug community is re-taking its place, for futures and learnings.

It is with great pride that we can say that, for the first time in 231 years, on 13th October, 2018, several hectares of land were returned at Blacktown in Sydney's west—returned to Dharug community, the traditional custodians of the majority of the Sydney Basin. These hectares are the site of what was the Black(s) Town Native Institution (BNI), which, together with the failed Parramatta Native Institution (PNI), was the start of the forced removal of Dharug and other non-white children from their parents. These removal practices later became known as "The Stolen Generations".[8] The process of seeking its return (in its most recent iteration) took five years of engagement with the NSW Government (Landcom) to achieve this goal, with all Dharug people on the committee being descendants of children who survived the PNI/BNI institutionalization. It is important to note that many didn't. This return is the first time that land has been given back to the traditional custodians of Sydney since early land grants to Nurrungingy, Colebee and Maria Lock were taken from their descendants by the Aboriginal Protection Board in 1920.[9] It is the first time that Dharug have been recognized as the rightful custodians of place. Aboriginal Land Councils can no longer say we don't exist. This is the leadership that will carry Dharug custodians into the future, bringing cultural authority, rather than bureaucratic power, which will be the hallmark of respected Aboriginal futures.[10] We are definitely taking our place and having our presence recognized. We are working towards having it respected for sustainable futures.

This chapter will introduce understandings of Indigenous presences, places and practices, through a Dharug lens, in order to privilege custodial concepts of caring for Country, connecting and belonging for sustainable outcomes when Dharug Ngurra is a cosmopolitan city. By weaving three examples (research, undergraduate teaching and learning and land return), it shows how Dharug Ngurra is activating, engaging and rejuvenating Dharug continuity through reflexive, experiential, network learning and empowerment. It argues that the educational praxis of Aboriginal yarning—as shared times/tellings, relevant to Country, and including Country—establishes the context that enables sustainable reflexivity for transformative learning and action when supported by experiential practices. Such a weaving is expressed as Goanna walking—a method that alerts us to our other-than-human partners-in-place. As we see massive bushfires and dried-up rivers across several states of Australia, which demonstrate humanity's neglect of landscapes, bio-diversities, and Indigenous cultural knowledges, it is argued here that reframing discourses and outcomes for sustainability must include futures grounded in Aboriginal custodial values. Climate change isn't just a rural phenomenon. It is a global phenomenon. It behoves everyone to care and act locally, even in cities.

Yarning-up place

Placing our values: Diverse ways of being, knowing and doing

Besides actually physically taking back Ngurra, through the BNI site return, Dharug are making place in the scholarship. Recent research engaged seven Dharug women (including myself as researcher-participant), yarning-up[11] our perspectives on Ngurra, the places of significance that represent continuity of culture, place and the important presences (as Ancestors).[12] These yarning sessions, held in-situ, relevant to the storying shared, brought to the literature both the diversity and the similarity of identities, values, and knowledges across the seven women's perspectives. In doing so, a picture of "Dharug-ness" became clear and was illustrated as a "Dharug Web of Interrelatedness": of belonging, caring, and connecting to presences, places and practices.[13]

What was uncovered through the research was the critical role "place" has within custodial cultural continuity. "Place" is one of those "wheelbarrow words" (Uncle Lex Dadd), that carries a lot of diverse meanings across diverse contexts. One of the key places for Dharug custodians is the place of values:

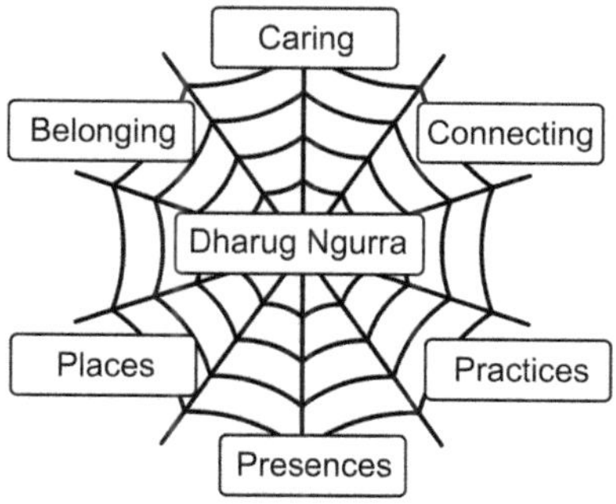

Figure 1.3 Dharug Web of Interrelatedness.
Jo Anne Rey, 2019

Spiritual, cultural and personal. This placement of values differs considerably to those that drive western humanism, as we see enacted on Dharug Ngurra and other large cities around the planet. Amber Sepie speaks to this difference in terms of "the re-ignition" of our consciousness of our human-more-than-human relationships.[14] Figure 1.4 below illustrates this difference.

What becomes obvious is that, at least since Plato's time in the Anglo-European patriarchal scholarship, and supported by Christian doctrines of the Earth as a resource for human use, humans have positioned themselves at the pinnacle of a species hierarchy, and intellectually at the centre of the Universe.[16] Privileging individualism and neo-liberalism for economic wealth places the individual as the self-referent centre of purpose and meaningfulness, while the environment is positioned secondary. In contrast, Dharug (and more broadly Aboriginal) Country-centric values place the collective wellbeing as relational to the wellbeing of Country, where the reciprocity of caring is relationally what underpins communities, families and the individual self.

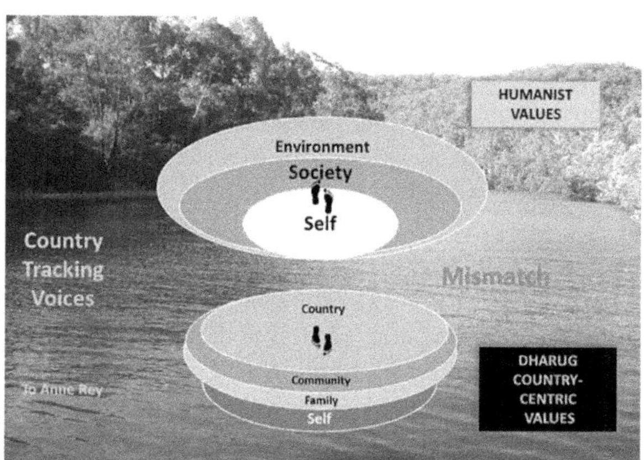

Figure 1.4 A Mismatch of Values: Human-centric vis a vis Country-centric.
Jo Anne Rey, 2019[15]

As Aunty Mary Graham[17] tells us:

> Aboriginal relationality—traditionally the foundation of the Law—is an elaborate, complex and refined system of social, moral, spiritual and community obligations that provided an ordered universe for people.

Core characteristics of this system are empathy/ethics, identity/place, autonomy and balance.[18] These characteristics are contrary to the core values expressed in the narratives of profit, economic success, multinational imperialism, globalisation, competition and individualism.

With such divergent values-placement underpinning human ontologies (the essence of being, beliefs and identifications), it is no surprise that Dharug Aboriginal ways of knowing and our epistemologies (relational knowledges through Law, language and Ngurra) are also divergent to those of the academy. Both Amber Sepie and Soren Larson, with Jay Johnson, articulate the significance of the differing conceptualizations of valid and invalid relationships with place and presences as coexistence, though coming at the topic from different directions—Sepie through storying, language and mythologies with other-than-humans, while Larson and Johnson address it through place.[19]

With a Country-centric way of being, our ways of knowing are also going to be centred in Ngurra and involve knowing through experiential, contextual learning for the pragmatic purpose of doing-as-caring. Such a praxis involves reciprocity with/as Ngurra. To clarify, when all being, knowing and doing is Country-referent and/or Country-refereed, such as shared yarning/telling-times in-situ, as was the case in Bawaka Country research, such transitions exemplify co-becoming—that is, becoming consciously together, even when Country is a city.[20] Consciously becoming together with Ngurra allows inclusive agency and brings reflexivity out of the human sphere and into the more-than-human. Arthur Middleton Young argued that this agency (which he called "monads" or "ghosts" that carry the "spark of life" behind the physics) was consciousness and underpinned what he called the "reflexive universe".[21] To this extent, Young's early work anticipates recent physics, post-qualitative and particularly post-humanist scholars such as Karen Barad, who speaks of the hauntologies that connect us as she crosses diverse disciplines, such as physics, philosophy, theatre and history.[22]

Changing places: Dharug knowings, learnings and doings as reflexive practice

Amanda Moffatt, Mary Ryan, and Georgina Barton argue that reflexivity is pivotal for transforming learning into action.[23] That is to say, it is in the mediations and adaptations that we make, between our internal (inner as psychological, emotional and spiritual states) and external circumstances (as material realities), that responsive action within context is possible. I argue that the educational praxis of Aboriginal yarning—as shared times/tellings, relevant to Country, and including Country, establishes the context that enables sustainable reflexivity for transformative learning and action when supported by experiential practices.[24] More than 65,000 years of Aboriginal ways of being, knowing and doing are the evidence of continuity embedded through place, and they allow room for intuitions. The place of intuition as a form of knowing is embedded in "Country-literacy" —reading place, space, time and agency—acquired through cultural transgenerational memory, in-situ learning and personal, intuitive response. Being Country-literate requires being "grown up" by Ngurra, and all within Her. Bringing Ngurra-centric ways of being, knowing and doing into the city, therefore has the potential to transform educational practices and outcomes for sustainable futures. Such is how Dharug community is taking its place and thus "co-becoming" as Ngurra.[25] It is also how caring for Country can bring change within Dharug Ngurra.

One significant way this is happening involves the Geography and Planning Department at Macquarie University (MQU), where third year undergraduate students and higher degree researchers are engaging with Dharug custodians and Ngurra near the Deerubbin River, of western Sydney.[26] With the Dharug consciousness of *yanama budyari gumada* (walking with good spirit), reflexive caring for Country is being undertaken through practices of Dharug-led cultural burns and Dharug-led culture camps at Yellomundee Regional Park. Through these experiential learning engagements, co-creating Country as reflexive learning-through-action brings to life the relational reciprocity that has underpinned sustainability on the continent for millennia. As Uncle Lex Dadd expresses it, "we're walking our Dreaming together now."[27]

The term "Dreaming" is particularly limited; however it has been adopted in English as shorthand to represent the cosmology of understanding that has underpinned 65,000 years of Aboriginal philosophies, storying and ways of knowing, being and doing. This chapter is not the place to expand extensively on this topic, but Larrakia (Northern Territory) Elder Bilawara Lee[28] helpfully differentiates the "Dreaming" from the "Dreamtime", in the following way:

> The Dreamtime and the Dreaming are not the same thing. Dreaming is the environment we live in, and it still exists today all around us. … We teach the knowledge deposited in the earth by the ancestral spirits. From the beginning of time, every event and every creative process that happened on Mother Earth effectively left behind a seed. This seed is seen as a memory that infused the earth after the event, just as a flower leaves behind a cloned copy of itself in the form of a discarded seed.
>
> All things formed during the Dreamtime carried with them a vibrational memory that holds the memories that birthed that place. … It [nature] carries the blueprint of the ancestral spirits whose actions helped shape the land. The memory stored within the land is what we refer to as 'the Dreaming', which represents the primordial sacredness of the earth.

"Walking our Dreaming together now", as such, recognizes the coexistence of making reflexive place, as a learning praxis, enacting and continuing culture,

Country and caring reciprocally, and demonstrates interconnectivity that fosters growth and heals disjunctions caused through colonization. Such a consciousness when shared disrupts the narratives of domination by settler-colonialism, over place, people, and ecologies, by uncovering and activating alternative practices, presenting alternative narratives, and displacing human power by embedding alternative presences as relational. Additionally, it resonates with Rey's Dharug *Web of Interconnectivity* (above), by enhancing belonging, caring, connecting with presences, place and practices. As burning Country has been one of the most significant sustainability practices that have actually shaped how the continent of Australia has become place (as a space of biodiversity), this practice is both the reflexivity and the reciprocity that makes us co-become more-than-human. It engages the mediations of the inner and outer circumstances—the giving and the receiving through experiential learning, which transforms through reflection into new understandings embedded in context.

Equally, significant place-changing is happening through the initiative taken by Macquarie University's Indigenous Studies Department under the leadership of Professor Bronwyn Carlson. Following from Rey's (2019) doctoral research, Professor Carlson saw the opportunity for the development of the 100 level unit: ABST1020 *Dharug Country: Presences, Places and People*—a Macquarie University (MQU) unit focused for the first time on Dharug custodian's perspectives, involving both Uncles and Aunties, and their relationships to places and presences on Country, cultural Law/Lore, and continuing practices. Making place within the University curriculum goes beyond recognition and acknowledgement of historic injustices. It recognizes and accepts the continuity of first cultural authority, knowledges and practices on the Dharug lands that were stolen.

The unit and, by extension, the University, accepts and includes current Dharug language that was previously forbidden up until the mid-20th century. It also recognizes the practice of continuing kin and spiritual relationship with other-than-humans, and thus, makes room for that continuity in the thinking and discourse as pedagogy in "the rock shelter" (classroom)—necessary for inclement weather and circumstances requiring technology. Bringing rock shelters and technology together, I will argue later in this chapter, is an example of what Rey calls "Goanna Walking",[29] or "third way" approaches.[30]

Through the emphasis on experiential learning, students go out to significant Dharug sites such as rock engraving platforms, shelters, and river sites with attached Dreaming storying. This practice exposes university students to learning through traditional custodial ways that have been practiced for thousands of years. Engaging with Ngurra (Country), its presences and spirituality becomes a continuity of practice, not only with Dharug students, but also other Indigenous and non-Indigenous class members. This enables changed futures, where values other than economic consumerism, individualism, and materialism are privileged for the sake of opening opportunities to re-think approaches for sustainable futures.

Taking the learning out of what Blair calls the "brick walls" and into Ngurra decolonizes systems that have privileged settler colonialism by re-establishing experiential relational place-based pedagogical approaches.[31] In a time when activism on the streets by predominantly young students, and associated older concerned citizens, is driven by motivations for sustainable futures, the necessity for opening hearts and minds to alternative ways of being, knowing and doing behoves higher education institutions to press pause on the "business as usual" models of engaging with/in the planet. Including Aboriginal understandings, which have been sustainably implemented for more than 65,000 years, is to recognize that knowledges in-place, knowledges in-context, and relationality are the underpinnings that do not destabilize organic and bio-diverse systems, but enhance and provide the sustenance for belonging, resilience and well-being. Using reflexive place-making pedagogies in a reflexive universe is to privilege alignments of energy, systems and understandings that bring webs of connection that foster growth, rather than disjuncture.

However, in order to bring five million hearts and minds into this reflexive "univers-(c)-ity" (Ngurra) we have to recognize the role of our relationality, our place of weaving between the constructions of the colonial mind, and that of the de-constructions of Ngurra. As mentioned earlier, that weaving across the landscapes of mentalities requires 'Goanna Walking'. Herein lies what makes these current transformative placements sustainable, relevant, reflexive, contextual and needs-based. In Goanna walking between the doing of the praxis out on Country, and the reflectiveness and criticality of the knowledges in the literatures to underpin it, walking between brings the power (of the western system) into context with the knowledge-authority of custodians; together, weaving a path for ethical futures.

Goanna walking towards sustainable and ethical futures

As Graham notes, for thousands of years, Aboriginal society was grounded in ethical relationality—"the foundation of the Law".[32] It recognised the sanctity of

> …the relationship between Land and human beings which, in turn, led to the fundamental principle of custodianship or a permanent, standing obligation to look after Land, society and social relations—the Law.

Such ethics were not human-centric, but recognized the sovereignty (rightful authority) of ecologies, the webs of connection, caring, belonging and practising, with human beings positioned within, not above them.[33] Graham also notes the place of protocols is critical as they are tools for negotiating respect and reciprocity, required for maintaining sustainable futures that are resilient and nurture wellbeing holistically. Implied within that is the role of an ethics-based education system that positions the citizen as well-educated when they can lead others for sustainable futures not for corporate profit and personal

Figure 1.5 Rosenberg Monitor Lizard.
Jo Anne Rey, 2017

success. It is suggested here that it takes "Goanna walking", as a method of ethically weaving between the colonizing dominant human-centric paradigm and the sustainable other-than-human ways of knowing, doing and being, that can bring such well-educated students to the task of creating sustainable futures.[34]

While in the thesis, "Goanna Walking" is recognised as the practice of weaving between two cultures: Western academic doctoral rigours and Dharug women's ways of yarning as educational transmission of transgenerational storying, understanding Goanna's role as one of "companion researcher" meant a place opened between responsibilities as research academic and responsibilities as a Dharug participant/community member. This was an ethical third-way place that had to be contemplated and reflexively engaged with. In the process, the mediations between inner (psychological, emotional, intuitive) and outer (individuals, group, cultures, places, presences, etc.) contexts wove transformational knowledges, which eventually became expressed through the creation of the Possum Skin Wrap, burnt with the map of the doctoral project journey (Figure 1.1).

As such, it was a creative and transformative process that involved Aboriginal ways of educational practice, that is, the experiential (the doing), contextual (thesis mapping through custodial knowledge transmission) and needs-based learning (a method that demonstrated the skills of possum work that have evolved through thousands of years when possum cloaks and other items, such as bags, head wear and blankets were worn and used for warmth, decoration and practicalities). "Goanna walking", as intercultural/third-way praxis, therefore, offers profound opportunities for non-Indigenous teachers and students to co-become with Ngurra and gain transformative and reflexive insights for sustainable futures. It also weaves a consciousness and ethics that recognizes ecologies-sovereignty. Developing curricula that incorporates Aboriginal-literacies could open pathways that enable coexistence that sustains all sentience and brings educational outcomes for Indigenous students into equity with their non-Indigenous peers.

Goanna walking: pedagogy as decolonization with/through/as ethical ecological relationships

The MQU undergraduate unit ABST1020: *Dharug Country: Presences, Place and People*, brings an ethics of ecologies-sovereignty to first year undergraduate students. By ecologies-sovereignty, I follow Legg, who positions sovereignty as "networked power".[35] As such, recognizing ecologies, as living networks, also recognizes living sovereignty and agency within place(s). Such recognition does not position human beings above other sentience but positions us within these ecologies as ethical relationship. It therefore follows Aboriginal ontologies, epistemologies and practices as Graham describes above through "Law" and at the same time brings opportunities for alternative ways of teaching, learning and doing with other-than-human presences. As such the unit enacts an ethical stance.[36]

For students to gain these insights, they are required to visit a range of Dharug sites of significance and use Dharug protocols of showing respect to Ancestors and Presences, engaging with the place and at that moment recording the experience, through photographs, video and written accounts. Some students utilize poetic responses. Some have used drones to provide alternative perspectives; some have simply sat and pondered the power of the light and the visitations of creatures. All have been required to contextualize the place within its colonized his-stories and pre-colonial Her-stories—as evident through scar-trees, rock engravings, middens, waterways and landscapes. As they have engaged fortnightly with the various sites (self-selected within a range offered), they have recounted the impact the visit has had on their consciousness of place, its agency, and their relationship with it. They have enacted ethical recognition of these ecologies-sovereignties.

Under the premise that connecting grows caring, and both grow a sense of belonging, the aim of this unit, therefore, is to change students' ways of knowing, being and doing, so that they can care for/as Country even though it is a city, through localized, thoughtful, respectful engagement. This is not to suggest that massive change will occur overnight, but rather, like the trickle of water through the crack in the brick wall, eventually daylight will shine, and an empathy for place, ecologies and other-than-humans will grow.[37]

At the same time engaging with/as Country, as experiential pedagogy, decomposes the human-centric privilege that separates human-other-than-human relationships, which have characterized much glorification of the academic and scientific objectification methodology over the past two thousand years.[38] As John Law establishes, the process of scientific investigations, particularly in laboratory settings, is based on the separation of elements into "relevant" and "not relevant" to the experimental (in contrast to experiential) method, while at the same time ignoring the subjective, worldview underpinnings present. As such it is fallacious to say it is in "objectivity" that the superiority of the scientific method resides.[39] Amber Sepie amplifies this disparity in process and methodology between colonial and Indigenous understandings when she states:

> Whatever shape it [knowledge] appears in, however, cosmologies (or origin stories) remain inherently linked to the pragmatic aspects of sustaining a human community, which means that interrogating cosmological matters becomes central to comprehending how a worldview actually works. The idea of reverence (the indigenous or eco-spiritual consideration for nature, or self-as-nature) and utilitarian principles are not mutually exclusive, but mutually dependent upon one another to be successful.[40]

Bringing these disparities together, and where the unit acts in a third, or Goanna walking way, is to weave oral, visual, experiential, place-based and written knowledges and practices together, thus establishing multi-modal pedagogical approaches, *relevant to context*, that situate and recognize humans with/in, rather than above, presences and places, and allow a place for multi-literacies. It is not to suggest that there is a constancy in human consciousness so that all actions are directed through Ngurra, as human-other-than-human engagement. Especially when Ngurra is a city, and we have all been colonized through mass education/indoctrination systems. However, when we are attentively engaging and present with/as Ngurra, we co-become as more-than-human. Such an approach embeds Indigenous ways of knowing, being and doing into pedagogical events and disrupts privileged practices such as written knowledges as an elitist hierarchical tool to discriminate against those who prefer other forms of knowledge delivery and transmission. As Van Toorn notes, "the connection between literacy and cultural 'advancement' is embedded in the English language in terms such as 'illiterate' and 'pre-literate'", and as such provide evidence of the disconnections, separations and hierarchies that have characterized educational practice by elites for thousands of years.[41]

However, as she also notes:

> …writing and literacy are never practised in a vacuum, … there is no such singular thing as 'literacy itself ', no single set of reading and writing practices that are inherently and invariably correct, but instead [there are] a multitude of ways to practise literacy. Literacy can therefore only be validly *examined in context, at particular sites* (my emphasis), rather than in abstract general terms.[42]

As such, bringing in-situ, experiential learning, into the academy is to bring another form of literacy to the notion of education and learning outcomes.

From an Indigenous perspective, reading Ngurra: its seasons, patterns, interactions, and consequences is a literacy that underpins 65,000 years of sustainability. Educating non-Indigenous peoples to become literate in reading, relating with and caring for/as Ngurra is at the heart of sustainable futures, especially when Country is a city. It also opens the University to changing their presences, places and practices for its sustainability and decolonized future.

Weaving Place(s) for Decolonized Futures

Weaving between past-presences

Many people, whether Indigenous, non-Indigenous from Australia, or other parts of the globe, struggle with closing the values gap between human-centricity and those centred in sustainable, relational reciprocity with Ngurra (as presences, places and people). Due to colonization, supported by a white supremist educational system, Stolen Generation practices, and genocidal intent on the part of settler-colonialism, many Indigenous people (young and old) are disconnected from their cultural places of belonging, their languages, their ways of knowing, doing and being, and their storied heritages.[43] As such many are seeking more understanding of what it is to belong, even though they might not articulate a sense of disconnection. Learning that belonging means to be part of the ecologies-sovereign web that involves caring and connecting to presences, places and practices is one of the student outcomes from the undergraduate Macquarie University ABST1020 unit.

Ngurra, as the largest city in Australia, with the highest human population, has a unique identity and place within the Australian Indigenous landscape.

Figure 1.6 MQU – Campus Mural by Dharug Artists: Leanne, Chris and Shey Tobin. Image: Jo Anne Rey, 2013.

As the first site of British take-over, Ngurra and Dharug people (along with their immediate neighbours) have the longest endurance of that cultural clash. We also have the longest opportunity for learning about the other, and the greatest experience on the continent of Australia, of surviving within such colonizing mentalities that drove people from their places of belonging, food sources and spiritual, ceremonial gatherings, in order to establish their own ways of knowing, doing and being. That is not to say all colonizing mentalities are enforced, or experienced, exactly the same—the experiences of Northern Territory peoples, for example, have been very different to those of Sydney. But as Tatz reminds us, the genocidal intent was evident across many frontiers of the continent.[44]

Nevertheless, when the largest number of Indigenous peoples today are living in urbanized places, and the majority of those are in Sydney's western suburbs, Dharug community have the rightful cultural authority and place to be the leaders in addressing cultural matters, including forming ways to undertake cultural practices, and obligations around caring for Ngurra and learning from Ngurra. It can also be argued, we have the greatest need. Bringing Indigenous peoples, non-Indigenous Australians, new arrivals from overseas and weaving them into the core fabric of Ngurra, so that they have a recognition of their place of belonging within Dharug Ngurra, is in the interests of everyone. Part of that journey is recognition of the values underpinning relational reciprocity, recognition of the obligations associated with reciprocity and recognising we are now "sharing our Dreaming together".[45] As such, learning for urban futures that are sustainable requires weaving with the wisdoms that can be shared, and weaving between the blockages that remain. In the context of pollution, climate change, and mass extinctions of bio-diversities, these values have never been more important.

Weaving webs of connection

Whatever the landscape, Ngurra teaches us about continuity and change and provides a culturally responsive pedagogy (CRP).[46] As Neil Harrison and Iliana Skrebneva found, as they investigated the feasibility of Country as pedagogical within the school sector, the search for a means to "measure success outside the annual *Closing the Gap*[47] targets and NAPLAN [National Assessment Program] results," involved only a very small number of studies because few have been undertaken in Australia.[48] However, they did find that "increased student engagement [was] demonstrated through indicators of belonging and identity..." and as such these factors become markers of successful pedagogy. Additionally, it was found that when Ngurra is recognised as teacher, "Country has the potential to displace the (colonial) authority of the teacher, ... and students and teachers are able to listen and observe the patterns of Country." To this extent "Country as pedagogical enacts the seasons, the direction of winds, tides, light and sun. It teaches through repetition and relationships, so the structure is already there for us to learn."[49]

Moving beyond the schooling sector and into the higher education sector, recognizing Ngurra as performative pedagogy, is being undertaken through the shared Dreaming webs of connections and continuing interactions that develop a sense of belonging and identity. As is being shown, when students engaged with Ngurra as pedagogical, they recognized a different way of being, knowing and learning, one that was not premised on learning from another person, but through their own experiences in-place.

As one student (AC) in the ABST1020 unit noted in her response to engaging with Ngurra at Brown's Waterhole:

> Walking to Brown's Waterhole this week was a beautiful experience intersecting a childlike sense of exploration, immense gratitude and reflection, thought-inspiring conversation, silence and beauty. It was wonderful to feel our shared congruence with presences, noticing how much more instinctual and organic our embodiment of the on-Country protocols has become. In theme with our topic—Futures, I have been asking myself: What kind of presence do I want to embody as an actor in the collective future? What responsibilities am I willing to acknowledge and accept as a carer for Country? And coming back to this concept of conscientious engagement with time, Ngurra this week has helped me remember the importance of time-spending (particularly doing-time and becoming-time), and how easily it can be forgotten that even small changes in time attention—such as a short walk to bush water, can drastically refocus what we value and quench mental wellbeing thirst.

Another student (MP)'s response reflected on what the pressures of urbanity for the site of Gore Reserve on Berry Island presents:

Figure 1.7 Brown's Waterhole. Jo Anne Rey, 2019

> I think Berry Island [Gammeraigal clan's Ngurra] and its immediate surroundings are the epitome of examples given to demonstrate the importance of these concepts. Here we have a small piece of land (Berry Island) covered in Indigenous history and pure, age-old flora that is by very definition encompassed by development, human indulgence, pollution, industry, modern housing, large-scale maritime operations and processing plants. Even as I walked along the Gadyan track I could see through the trees an enormous maritime vessel docked not even 100 metres from the Island's shore. It is thus of paramount importance that the never-ending growth of anthropocentricism manifest in urbanisation does not result in encroachment onto Berry Island.

And on Bidgigal Reserve a third student (SW) writes:

> Roaming the country always seems to give me a sense of belonging, that I have found something I didn't know I was searching for. It has the ability to create its own magic to transform you from the now, and to take you back to a time that we can never reach. Only through exploring Ngurra may we begin to piece it all together, through interpretation and one's own experience may it all begin to make sense, why the rock shelters why the locations of their sites, why they chose to live so peacefully amongst this beautiful land and just what it was to be indigenous amongst Ngurra.
>
> It is almost an ancient feeling, being amongst our true environment, being exposed to the elements and at the mercy of our own instincts, what it truly means to be alive. Its more than just a state of being, it's a way of life, surviving each day, lasting the everlasting, defying death, all for the reasons of existence.

Each of these examples shows the effect that engagement with Ngurra in its variety has, and how it acts to heighten a consciousness of caring that keeps connections open, that keeps humans and those bio-diversities alive to sustain futures. These connections are the basis of the ecologies-sovereignty web that continues our Dreaming, our Law and our belonging. They also keep everyone in Ngurra connected.

Weaving webs of being, knowing and doing

Being heard, being seen and being trusted are the threads that can weave sustainable reciprocal relationships. Dharug community are finding ways of being heard. Examples include local Council initiated activities, undertaking Welcomes to Country for organizations, educational institutions, and through community artistic events.

Similarly, through participation in the celebration of Indigenous cultural events, such as during NAIDOC week, and/or artistic endeavours, such as the artist's camps on the (now returned) BNI site (organised with Blacktown

Council, Museum of Contemporary Art and Dharug community), or the Biennale in 2020, where the BNI site is registered as an Artist, Dharug community are being seen and are regaining their place and their cultural authority. It is through the relationships that are made during these engagements, that Dharug people are being recognized and being trusted, and reciprocally, those working, engaging with and entrusting Dharug community, are being trusted *by* Dharug people.

Knowing presences, place and people: Building webs of strategic alliance

Through these engaging and entrusting relationships, growing sustainable reciprocity is being enlivened. Whether it is through the cultural camps happening at Yallomundee (Penrith), undergraduate student out-on-Country experiences (ABST1020), or participation in award winning public partnership engagements, Dharug community members know they are making a difference, and people are making a place for them. Getting on with the doing, getting on with the being, getting on with making change, builds a knowledge base founded in alliances. Such has been cultural practice for millennia. Not through competition, but through strategic alliance formations that build strong communities, that support strong families, and that raise strong individuals: strong in belonging, strong in caring and strong in connecting, to the presences, the places and the practices.

Doing for sustainable futures: Weaving purpose for perpetuity

However, it cannot be underestimated the significance of the historic return, on 13th October, 2018, of the Blacktown Native Institution (BNI) site—not to Aboriginal Land Councils—but to Dharug custodians, as a site to be held in trust for all Dharug community. The site is situated in urban-Aboriginal and Torres Strait Islander heartland, where hundreds of families are working and raising their children.

For Dharug custodians, it is now impossible for Land Councils to say Dharug custodians don't count; don't have a voice; don't have a right to be heard—even that we don't exist! For Dharug custodians, the five years journey taken by the Interim Working Committee, formed to see the return of the site to custodians, has created a benchmark for successful engagement by other organizations working within the Indigenous space. To reach that benchmark it took walking and weaving (Goanna-like) between naysayers from within Aboriginal spaces, local and state bureaucrats worried about timelines and bottom lines and the dogged dedication of the handful who were persistently, quietly engaging in trust-building, knowledge and capacity-building, underpinned by values of honesty, sharing, caring and connecting.

That final handful at first glance would seem an unlikely alliance. Firstly, the NSW Government (Landcom) was the owner of the title. As the government, they represent the power that has colonized Australia and originally removed the children and began and perpetuated the Stolen Generations. Secondly GHD were the consultants employed to bring the parties together. Thirdly, Dharug community members, who are the descendants of some of those little children who had been the inmates that had survived their institutionalization, were the representative recipients.

Together open communications channels were forged. This was no easy task, and fixed views had to be adjusted. This took time, and it took the right people being in place. Over the years several participants changed. There were lull periods where people didn't quite fit. But there was always a core group that maintained the continuity—of vision, of storying, of collective engagement. When changes were made new relationships had to be formed. Eventually the right people, at the right time came together. The heart of the web of connections—an ecology of sovereignty was formed. As relationships and trust grew a sense of shared belonging arose—belonging to the vision of the purpose: to return to custodians, for the first time in around 200 years a place to keep in trust for community; a place to practice culture; a place to remember Ancestors who had either died or survived the mentalities of the colonial day; a keeping place for knowledges, artefacts, and storying, and a place to build strong community, strong families and strong individuals that have a strong sense of belonging, caring and connecting to Ngurra—its presences, places and practices, for the benefit of all sentient beings today and in the future.

Eventually, a timeline had to be adhered to. This grounded us, and focussed us, with the pace increasing rapidly as required steps were put in place. While the time span for a leisurely Goanna walk had been undertaken for several years, when the realisation came that the handover was actually going to happen, there was little wriggle room, and a directed path was put in place.

The committee had to be transformed into a legal entity. Funding for survival had to be found. Media and communications, inside and outside of community, had to be organized. Dates that appealed to politicians had to be found. The handover event had to be prepared. A corroboree was arranged. Notifications and invitations had to be sent. Logistics and practicalities of the event were set in place.

And it happened: A smoking ceremony to welcome everyone. A corroboree with dancers and the story of the site, through the naming of the children, was told. The official handover and speeches were made. Gift giving from Dharug community as reciprocity took place.

And across the site, it rained, and the kangaroos returned through the mists in the distance. This was our Dharug ecological-sovereignty being enacted.

A year later, DSMG/GHD/LANDCOM were awarded the 2019 Indigenous International Association of Public Participation (IAp2) Award.

Becoming place(s) for learning futures

In the face of a changing climate and massive extinctions of ecologies, I have woven presences, places and practices to call for recognition of the urgent need

Figure 1.8 Blacktown Native Institution (BNI) Site Handback 13.10.2018.
Jo Anne Rey, 2018

Figure 1.9 BNI Handback – Corroboree.
Jo Anne Rey, 2018

to redress the neglect of values that underpin meaningful custodianship of Country for the benefit of sustainable futures. Recognition of ecological sovereignty requires a re-positioning of humanity, away from seeing human beings at the top of a hierarchy of species and towards a respect for humanity's place within and dependent upon those webs of connection. When Country is a city, of millions of people, in order for a suitable response to this need to be apprehended, a change in approach is required. It has been proposed that change involves bringing Aboriginal values and ways of knowing, being and doing to the broader population, because the original custodians have been caring for Ngurra for tens of thousands of years. It is arrogance and/or

Figure 1.10 BNI Handback Corroboree 13.10.2018. "Petticoat Dance" – Jannawi and Dharug Dancers.
Jo Anne Rey, 2018

ignorance to disrespect this fact. Humanity can no longer afford a "business as usual" approach.

Through a localized focus, it is possible to privilege a consciousness that involves connecting to places, caring for those places and opening the opportunity to receive the benefits of belonging to those places, as Ngurra gives back to us reciprocally, through better health, resilience and sustainable wellbeing. Aboriginal ways have always involved the collective caring for Country. Collectively caring has involved strengthening communities. Strong communities support their families and strong families raise children who are strong in belonging, caring and connecting. Utilizing custodial ways of knowing involves custodial ways of practising culture that has been the foundation—the Law and the Lore—of continuity across 65,000 years interwoven across the continent. Such is the definition of sustainability at the human level, bringing about sustainability at the ecological level for sustainable futures. It is argued here that reframing discourses and outcomes for sustainability must include futures grounded in Aboriginal custodial values.

Efforts within the broader education system have begun, but as we see climate change impacting and laying waste through out-of-control bushfires to vast tracts of land, destroying wildlife and ecologies of flora and fauna, there is a greater imperative to reconsider humanity's options.[50] Narratives promoting consumerism, with economics as the God of wellbeing must change, and at least be brought into balance. Overproduction of materials, plastics, and non-recyclables are commonly stated culprits, as we see media footage of vast tracts of garbage in our oceans, our Great Barrier Reef under threat, rivers running dry from corrupt water management, and watery eyes are raised as Sydney chokes from surrounding bushfires' smoke, with the question: When is enough, enough?

Sustainable futures involve and most urgently impact the young people, and it is the young people who are currently carrying the burden of consciousness for the wellbeing of the planet and future generations of humanity. As protests, led by the young, who are making headlines globally as they recognize the human vulnerability to catastrophic consequences that climate change carries, many "Olders" in political positions of power sit back and privilege their continued wealth formation, utilization of resources (such as coal extractions, water licenses that privilege the mass producers rather than small farmers, and degradation of the marine stocks through over-fishing) at the cost of long term systems desperately needing attention.

Education for sustainable futures requires changing our values, and teaching our youngest to connect with their places, their heritages and their communities first, before seeking dreams over the horizons. Unless change happens, horizons will not be destinations wanting to be experienced, but boundaries keeping out threats, while locals will struggle to save the little that is left. Already, the global pandemic of coronavirus is making us awake to how interconnected we are. Between the fears of untenable futures, and urgency for responses to the realities of today, I suggest Indigenous knowledges have found their time and place in the broader narrative. Now it is up to everyone to recognize that they matter. Walking between, and weaving with Goanna,

Figure 1.11 Presences, Places and Practices – Healing Country.
Jo Anne Rey, 2017

allows other-than-human space to be negotiated, a more-than-human solution to become.

Notes

1 Dr Jo Anne Rey is a Dharug community member who is woven into Sydney's Dharug (Randall) and colonial (Squires) storying through ancestors, presences and places, particularly across Wallumattagal, Garigal and Cattai Countries. She has recently completed her doctoral thesis: *Country Tracking Voices: Dharug Women's Perspectives on Presences, Places and Practices* (Macquarie University, Wallumattagal Country, 2019). It attends to the voices and perspectives of seven Dharug 'sistas', yarned from their significant places of connection, about Dharug continuity when Country is a colonized, cosmopolitan city. She has a professional background as a teacher-librarian, a M. Ed. (International) (Monash, 2008) and is an historical-fiction author (*The Sarsaparilla Souvenir*, Xlibris, 2005). She is a foundational Board member of the Dharug Strategic Management Group (DSMG) and sees her participation as one way she cares for Country, Ancestors and Dharug community.

2 Kerry Arabena, *Becoming Indigenous to the Universe: Reflections on Living Systems, Indigeneity and Citizenship* (North Melbourne: Australian Scholarly Publishing, 2015), 10.

3 'Goanna' is the Australian English term for monitor lizard.

4 Jo Anne Rey, *Country Tracking Voices: Dharug Women's Perspectives on Presences, Places and Practices* (Sydney, Australia: Macquarie University, 2019).

5 Homi Bhabha, "The third space," *Identity: Community, Culture, Difference*, ed. Jonathan Rutherford (London: Lawrence & Wishart, 1990), 207–221.

6 Soren C. Larsen and Jay T. Johnson, *Being Together in Place: Indigenous Coexistence in a More Than Human World* (Minneapolis: University of Minneapolis Press, 2017).

7 Jens Korff, 2019, "Creative spirits," https://www.creativespirits.info/aboriginalculture/people/aboriginal-population-in-australia.

8 For background information see: Australian Commonwealth Government, *Bringing them Home: National Inquiry into the Separation of Aboriginal and Torres Strait Islander Children from their Families*. Canberra, ACT, 1997.

More recent information can be found from: *Grandmothers Against Removal*http://stopstolengenerations.com.au/2016/05

9 Australian Dictionary of Biography, http://adb.anu.edu.au/biography/lock-maria

10 Mary Graham, "Aboriginal notions of relationality and positionalism: A reply to Weber". *Global Discourse* 4 (1): 17–22, 2014. doi:10.1080/23269995.2014.895931

11 'Yarning-up': Aboriginal English meaning sharing stories and times.

12 Rey, *Country Tracking Voices: Dharug Women's Perspectives on Presences, Places and Practices*.

13 Rey, *Country Tracking Voices: Dharug Women's Perspectives on Presences, Places and Practices*, p. 318.

14 Amber Sepie, "More than stories, more than myths: animal/human/nature(s) in traditional ecological worldviews," *Humanities*. 6 (4): 2, 2017.

15 Rey, *Country Tracking Voices: Dharug Women's Perspectives on Presences, Places and Practices*, 40.

16 Val Plumwood, *Feminism and the Mastery of Nature*. London: Routledge, 1993.

17 Graham, *Aboriginal Notions of Relationality and Positionalism: A Reply to Weber*, 17.

18 Graham, *Aboriginal Notions of Relationality and Positionalism: A Reply to Weber*.

19 Sepie, *More than Stories, More than Myths: Animal/Human/Nature(s) in Traditional Ecological Worldviews*; Soren C. Larsen and Jay T. Johnson. *Being Together in Place: Indigenous Coexistence in a More Than Human World*. Minneapolis: University of Minneapolis Press, 2017.

20 Bawaka Country, Wright, S, Suchet-Pearson, S, Lloyd, K, Burarrwanga, L, and Ganambarr, R. *Co-becoming Bawaka: Towards a Relational Understanding of Place/Space.* Progress in Human Geography, 2015.
21 Arthur Middleton Young. *The Reflexive Universe: The Evolution of Consciousness* New York, NY: Delacorte Press, 1976.
22 Karen Barad, "Quantum entanglements and hauntological relations of inheritance: Dis/continuities, spacetime enfoldings, and justice-to-come." *Derrida Today, 3* (2): 240–268, 2016. doi:10.3366/drt.2010.0206
23 Amanda Moffatt, Mary Ryan, and Georgina Barton, "Reflexivity and self-care for creative facilitators: Stepping outside the circle," *Studies in Continuing Education* 38, (1): 29–46, 2015. doi:10.1080/0158037X.2015.1005067
24 Rey, *Country Tracking Voices: Dharug Women's Perspectives on Presences, Places and Practices,* 347–348.
25 Bawaka, Laklak Burarrwanga, Ritjilili Ganambarr, Merrkiyawuy Ganambarr-Stubbs, Banbapuy Ganambarr, Djawundil Maymuru, Sarah Wright, Sandie Suchet-Pearson, Kate Lloyd, and Jill Sweeney, "Co-becoming time/s," *Methodological Challenges in Nature-Culture and Environmental History Research,* eds. Jocelyn Thorpe, Stephanie Rutherford and L. Anders Sandberg. New York: Routledge, 81–92, 2016.
26 Darug Ngurra, Lexodious Dadd, Paul Glass, Rebecca Scott, Marnie Graham, Sara Judge, Paul Hodge, and Sandie Suchet-Pearson. "Yanama budyari gumada: reframing the urban to care as Darug Country in western Sydney." *Australian Geographer* 50 (3): 279–293, 2019. doi:10.1080/00049182.2019.1601150.
27 Ngurra, et al., *Yanama Budyari Gumada,* 280.
28 For more on the word "Dreaming" see Bilawara Lee, *Healing from the Dilly Bag.* Philadelphia, Pennsylvania: Xlibris: 12, 2013.
29 Jo Anne Rey, "Dharug custodial leadership: Uncovering Country in the city". *World Indigenous Nations Higher Education Journal,* 1: 56–66, 2019; Jo Anne Rey and Neil Harrison, "Sydney as an Indigenous place: 'Goanna walking' brings people together." *AlterNative: An International Journal of Indigenous Peoples,* 2019. doi:10.1177/1177180117751930; Rey, *Country Tracking Voices: Dharug Women's Perspectives on Presences, Places and Practices.*
30 Homi Bhabha. "The third space". *Identity: Community, Culture, Difference,* Jonathan Rutherford, ed. London: Lawrence & Wishart, 207–221, 1990.

Pat Dudgeon and John Fielder. "Third spaces within tertiary places: Indigenous Australian studies." *Journal of Community & Applied Social Psychology* 16, (5): 396–409, 2006. doi:10.1002/casp.883.
31 Nerida Blair. *Privileging Australian Indigenous Knowledge: Sweet Potatoes, Spiders, Waterlilys and Brick Walls* Champaign, Illinois: Common Ground, 2015.
32 Graham, *Aboriginal Notions of Relationality and Positionalism,* 18.
33 Ecological sovereignty is a fluid field of contestation, particularly but not only, in law. See Peter Boulot, "A new legal paradigm: Towards a jurisprudence based on ecological sovereignty." *Macquarie Journal of International and Comparative Environmental Law* 8, (1): 1–15, 2012.
34 Bawaka Country, Burarrwanga, L., Lloyd, K., Wright, S., Suchet-Pearson, S. Lloyd, K., Ganambarr, R., Ganambarr-Stubbs, M., Ganambarr, B., Maymuru, D. "Working with and learning from Country: decentring human authority." *Cultural Geographies.* doi:10.1177/1474474014539248
35 Emily Legg. *Writing Orality: Claiming Rhetorical Sovereignty Within Ecologies of Rhetorics.* Jennifer Bay, Samantha Blackmon and Thomas Rickert, eds. ProQuest Dissertations Publishing, 6, 2011.
36 Graham, *Aboriginal Notions of Relationality and Positionalism.*
37 Rey, *Country Tracking Voices: Dharug Women's Perspectives on Presences, Places and Practices.*
38 Ngurra, et al., *Yanama Budyari Gumada,* Bawaka, et. al. Co-becoming times.

39 John Law. *After Method: Mess in Social Science Research*. New York: Routledge, 2004.
40 Sepie, *More than Stories, More than Myths: Animal/Human/Nature(s) in Traditional Ecological Worldviews*, 15.
41 Penny van Toorn. *Writing Never Arrives Naked: Early Aboriginal Cultures of Writing in Australia*. Canberra: Aboriginal Studies Press, 9, 2006..
42 van Toorn, *Writing Never Arrives Naked*, 9.
43 Bronwyn Carlson. "Politics of identity: Who counts as Aboriginal today?" Sydney: University of New South Wales, 2011.
44 Colin Tatz. *Australia's Unthinkable Genocide*. (Bloomington, USA: Xlibris, 2017).
45 Ngurra et al. *Yanama Budyari Gumada*, 280.
46 Geneva Gay. "Culturally responsive teaching: Theory research and practice" *Multicultural Education Series*. New York: New York Teachers College, 2010.
47 Megan Davis, "Closing the gap in Indigenous disadvantage: A trajectory of Indigenous inequality in Australia." *Georgetown Journal of International Affairs*, 16 (1), 2015.
Reid, Allan, "Closing the achievement gap': Will the national education agenda be a help or hindrance?" *Teaching History* 45 (4): 49–58, 2011.
48 Neil Harrison and Illiana Skrebneva. "Country as pedagogical: enacting an Australian foundation for culturally responsive pedagogy." *Journal of Curriculum Studies*. doi:10.1080/00220272.2019.1641843, 2.
49 Neil Harrison and Illiana Skrebneva. "Country as pedagogical: enacting an Australian foundation for culturally responsive pedagogy." *Journal of Curriculum Studies*. doi:10.1080/00220272.2019.1641843, 2.
50 Neil Harrison, Susan Page, Leanne Tobin. "Art has a place: Country as a teacher in the city." *Educational Philosophy and Theory*, 48, 13: 1321–1335, 2016.
Ngurra et al. *Yanama Budyari Gumada*, Gus Worby, Rigney, Lester-Irabinna Tur, Simone Ulalka. *Where Salt and Fresh Waters Meet: Reconciliation and Change in Education*. Australian Cultural History 28 (2): 201–224, 2010.

References

Arabena, Kerry. *Becoming Indigenous to the Universe: Reflections on Living Systems, Indigeneity and Citizenship*. North Melbourne: Australian Scholarly Publishing, 2015.

Australia, Commonwealth of. "Bringing them home: National inquiry into the separation of Aboriginal and Torres Strait Islander children from their families." Canberra, 1997.

Australian Dictionary of Biography, http://adb.anu.edu.au/biography/lock-maria.

Barad, Karen. "Quantum entanglements and hauntological relations of inheritance: Dis/continuities, spacetime enfoldings, and justice-to-come". *Derrida Today*, 3 (2): 240–268, doi:2010. doi:10.3366/drt.2010.0206

Bawaka, Burarrwanga, Laklak, Ganambarr, Ritjilili, Ganambarr-Stubbs, Merrkiyawuy, Ganambarr, Banbapuy, Maymuru, Djawundil, Wright, Sarah, Suchet-Pearson, Sandie, Lloyd, Kate and Sweeney, Jill. "Co-becoming time/s". *Methodological Challenges in Nature-Culture and Environmental History Research*. Jocelyn Thorpe, Stephanie Rutherford and L. Anders Sandberg, eds. New York: Routledge, 2016.

Bawaka, Wright, Sarah, Suchet-Pearson, Sandie, Lloyd, Kate, Burarrwanga, Laklak, and Ganambarr, Ritjilili. "Co-becoming Bawaka: Towards a relational understanding of place/space". *Progress in Human Geography*, 2015.

Bawaka Country, Burarrwanga, L., Lloyd, K., Wright, S., Suchet-Pearson, S.Lloyd, K., Ganambarr, R., Ganambarr-Stubbs, M., Ganambarr, B., and Maymuru, D. "Working with and learning from Country: decentring human author-ity." *Cultural Geographies*, 2014. doi:10.1177/1474474014539248

Bhabha, Homi. "The third space," *Identity: Community, Culture, Difference*. Jonathan Rutherford, ed. London: Lawrence & Wishart: 207–221, 1990.
Blair, Nerida. *Privileging Australian Indigenous Knowledge: Sweet potatoes, spiders, waterlilys and Brick Walls*. Champaig: Common Ground, 2015.
Boulot, Peter. "A New Legal Paradigm: Towards a Jurisprudence Based on Ecological Sovereignty." *Macquarie Journal of International and Comparative Environmental Law* 8, (1): 1–15, 2012.
Carlson, Bronwyn. *Politics of Identity: Who Counts as Aboriginal Today?*Sydney: University of New South Wales, 2011.
Davis, Megan. "Closing the gap in Indigenous disadvantage: A trajectory of Indigenous inequality in Australia." *Georgetown Journal of International Affairs*, 16 (1), 2015.
Dudgeon, Pat, and John Fielder. "Third spaces within tertiary places: Indigenous Australian studies." *Journal of Community & Applied Social Psychology* 16 (5): 396–409, 2006. doi:10.1002/casp.883
Gay, Geneva. "Culturally responsive teaching: Theory research and practice" *Multicultural Education Series*. New York: New York Teachers College, 2010.
Graham, Mary. "Aboriginal notions of relationality and positionalism: a reply to Weber." *Global Discourse* 4 (1): 17–22. 2014. doi:10.1080/23269995.2014.895931.
"Grandmothers against removal." Wordpress, http://stopstolengenerations.com.au/2016/05.
Harrison, Neil, and Skrebneva, Iliana. "Country as pedagogical: enacting an Australian foundation for culturally responsive pedagogy." *Journal of Curriculum Studies*, 2019. doi:10.1080/00220272.2019.1641843.
Harrison, Neil, Susan Page, and Leanne Tobin, "Art has a place: Country as a teacher in the city." *Educational Philosophy and Theory*, 48 (13): 1321–1335, 2016.
Korff, Jens. "Creative spirits." 2019. https://www.creativespirits.info/aboriginalculture/people/aboriginal-population-in-australia.
Larsen, Soren C. and Jay T. Johnson. *Being Together in Place: Indigenous Coexistence in a More Than Human World*. Minneapolis: University of Minneapolis Press, 2017.
Lee, Bilawara. *Healing from the Dilly Bag*. Philadelphia, Pennsylvania: Xlibris, 2013.
Legg, Emily, Jennifer Bay, Samantha Blackmon and Thomas Rickert. "Writing orality: Claiming rhetorical sovereignty within ecologies of rhetorics". Masters Thesis, Purdue University, West Lafayette, Indiana, USA: ProQuest Dissertations Publishing, 2011.
Moffatt, Amanda, Mary Ryan, and Georgina Barton. "Reflexivity and self-care for creative facilitators: Stepping outside the circle." *Studies in Continuing Education*, 38 (1): 29–46, 2016. doi:10.1080/0158037X.2015.1005067.
Ngurra, Darug, Lexodious Dadd, Paul Glass, Rebecca Scott, Marnie Graham, Sara Judge, Paul Hodge, and Sandie Suchet-Pearson. "Yanama budyari gumada: reframing the urban to care as Darug Country in western Sydney." *Australian Geographer*, 50 (3): 279–293, 2019. doi:10.1080/00049182.2019.1601150.
Plumwood, Val. *Feminism and the Mastery of Nature*. London: Routledge, 1993.
Reid, Allan. "Closing the achievement gap": will the national education agenda be a help or hindrance? *Teaching History*. 45 (4): 49–58, 2011.
Rey, Jo. "Country tracking voices: Dharug women's perspectives on presences, places and practices," Sydney, Australia: Macquarie University, 2019.
Rey, Jo. "Dharug custodial leadership: uncovering Country in the city." *World Indigenous Nations Higher Education Journal*, (1): 56–66, 2019.

Rey, Jo, and Neil Harrison. "Sydney as an Indigenous place: 'Goanna walking' brings people together." *AlterNative: An International Journal of Indigenous Peoples*. 2018. doi:10.1177/1177180117751930.

Sepie, Amber. "More than stories, more than myths: Animal/human/nature(s) in traditional ecological worldviews." *Humanities*, 6 (4): 2017.

Tatz, Colin. *Australia's Unthinkable Genocide*. Bloomington: Xlibris, 2017.

van Toorn, Penny. *Writing Never Arrives Naked: Early Aboriginal Cultures of Writing in Australia*. Canberra: Aboriginal Studies Press, 2006.

2 Kichwa stories of future(s)

Narratives for otherwise good living

Gioconda Coello

In Kichwa and Quechua thought,[1] the future is neither a time to arrive nor a time that one must face; rather, it is behind us and returns again.[2] Kichwa notions of time and anticipation of time relate to reading what is coming into being together through relationships to the land and people. This chapter will explore those relationships as they participate in storying and giving continuation to knowledge and ways of learning about a good life, and disrupting bureaucratized notions of "good living." In the past decade, Ecuadorian public policies and projects have risen for planning the nation's future to achieve a society of "good living."[3] These proposals claimed to use Kichwa epistemology to center politics, economy, and education on the human being and a harmonious relationship with nature. However, the theorization, planning, and implementation of these proposals follow developmental logics rather than Kichwa ideas of the human-nature relationship and of a good future. This chapter attempts to disrupt the narrative these projects produce about "good living" and open conversations about a "good living" for the future(s) embedded in Kichwa stories about the world, the land, dreams and everyday living.

Official and bureaucratic discourses use the Kichwa words Sumak Kawsay (SK) and its translation to Spanish Buen Vivir (BV) as notions of "good living" supposedly coming from Andean Indigenous wisdom.[4] These terms now defined in the constitution and multiple political discourses and projects are, importantly, used in the government planning of a "good life." In so doing, that planning establishes who is included in the making of the future of the country, who is not, and who compromises the desired future in the present. This planning is not unlike other practices of national security in which a disclosure of the future, in this case a desired society of "good-living," creates a civilized, "legitimate" population and an uncivilized, illegitimate, risky population.[5] Education thus becomes a means by which to transmit values seemingly "ancestral" to model citizens before that risky population endanger the ideal "good-life for all."

In this chapter I contrast the logic underpinning governmental stories about "good living" in section two with the logic in Kichwa stories about future and "good living" in section three. I argue that Kichwa ways of learning about future(s) to live well disrupt extant bureaucratized conceptions of "good living." Drawing on Sylvia Wynter's ideas of the *homo narrata*[6] I propose stories to be the narration of truths from diverse ontological possibilities according to the ways of life and thinking of the (human and more-than-human) storytellers.

The first section discusses academic stories written by authors from the Americas that provoked my questioning of time and stories of future(s) which I, as a descendant of peoples of the Andes and its *ch'ixi*[7] society and being an academic migrant, kept in mind while learning more about Kichwa future(s). The following academic stories explore being human, time, place, and pedagogy and are cited as the genre requires. The chapter counterposes academic stories with Kichwa stories to engage with the epistemological abyss[8] implied in any conversation about "good living." Doing so blurs the divisions that European and American critical traditions have drawn between their ways of knowing and other ways of knowing particularly Indigenous ways of knowing as shared by Indigenous thinkers outside academia. Counterposing in this text refers to the effort to bring together different bodies of thought to undo the singularity imposed on "future" from multiple perspectives. As previously indicated, I draw on Wynter's *homo narrata*, on Gerald Vizenor's survivance, and on Kastuve Roy's pedagogical encounters to highlight the creation of stories through learnings and curricula coming from diverse social locales. The second section presents the stories of the Ecuadorian government about the meaning of Sumak Kawsay and Buen Vivir and their use of "ancestral" knowledge to plan development in ways marked by coloniality. The third and largest section comes from stories of friends and instructors shared with me in Ecuador. It focuses on Kichwa oral stories about future(s) and learnings for living well which interrupt "good living" as life planning that closes the multiplicity of future(s) and singularizes the ways of knowing it, learning about it, and creating it. The stories and the multiplicity of future(s) point towards a kind of curriculum that is not necessarily planned but conjured.

Storied future(s): Continuance of signs, presences, and encounters

In this section, I propose that future(s) are stories that have not yet taken place. Those stories come into being by a narrator who reads signs collectively produced. The collective is people understood somewhat as runa, in Kichwa language. Runa, which can be translated as a person or people, describes all beings who are familiar and pertain to the Kichwa life. Trees and animals are also runa when they belong as actors in the world of the Kichwa speaker and are therefore people. There are more nuances to the category of people when beings are unfamiliar, but these are beyond the scope of this chapter. What I attempt here is to make clear my understanding of a storyteller who can be more-than-human. The signs used by human story tellers to weave narratives might as well be a part of other people's stories. Wynter asserts that human beings are a *homos narrata*, a "hybrid-auto-instituting-languaging-storytelling species."[9] She argues that apart from the biological existence humans exist through words and stories. Stories are formative of the possibilities of being and the possibilities of the world we make. She urges academics to understand the functioning of the world-systemic societal order not as a given but as a sociogenic production. The Fanonian concept of sociogeny is a transcultural space that culturally articulates a "sense of self."[10] Sociogeny sets "man" free of oppressive, universalizing, and singularizing descriptive statements and stories of identity on what it means to be human, which are traceable to the Renaissance and informed by social Darwinism.[11] The same is true about time and histories. Times can be freed from singularizing progressing time that binds pasts, presents and futures to cause-effect chains. Histories can be freed from stories that dismiss their construction and the multiple possibilities of story telling that survive even if untold.

Indigenous stories are a way in which important untold histories survive and make present many senses of self. Here survivance has the quality of Gerald Vizenor's survivance. In his words, "the practices of survivance create an active presence, more than the instincts of survival, function, or subsistence."[12] Indigenous stories do not simply allow the continuation of knowledge but conjure the presence of untold possibilities and relations among people and with the land. Survivance is also a practice of narration that makes presence which "arises from experiences in the natural world, by the turn of seasons, by sudden storms, by migration of cranes by the ventures of tender lady's slippers, by chance of moths overnight, by unruly mosquitoes, and by the favor of spirits in the water, rimy sumac, wild rice, thunder in the ice, bear, beaver, and faces in the stone."[13] The survivance provoked by Indigenous stories thus weaves presences through multiple selves encountering each other. Vizenor focuses on Native American stories; however Kichwa stories too weave presences and selves as in them the world and people emerge relationally. Eduardo Kohn's *How Forest Think*[14] is an important contemporary work that explores this continuation of selves or survivance.

Khon regards the Kichwa future as a realm that is inhabited and produced by semiotic selves. Human and more-than-human selves, he notes, are outcomes and starting points of the process of interpretation of signs that produce future selves. In this sense, selves emerge relationally and become visible in the making of what will be. Visible selves and possibilities of the future survive the silence of possibilities that persist even if invisible. As Kohn explains, "[l]ife grows in relationship to that which it is not,"[15] in which relationship is a real and continuous link that is necessary for becoming a self. This is so because it is an outcome of processes that maintain a certain closure, or individuality, but always respond to the surrounding world.[16] Like Vizenor points out, life survives through the presence of multiple beings related to each other. Within those relationships, people narrate selves, thus placing them and making them possible and visible within a story.

The future and the past, in Khon's analysis, occupy the same realm located inside the forest as a natural and supernatural semiotic place. Reading the signs in the forest brings about possibilities that follow a logic that makes the future "not reducible to that which will have happened"[17] and able to house all the possibilities of the past.[18] That is, to live and make future are processes that are bound to (human and more-than-human) selves' ability to persist as an "I." To this extent, the future is multiple (hereafter future[s]) as there are multiple and simultaneous narratives about what will happen which depend on the interwoven presences of multiple beings as well as on how readers and narrators story those presences.

Reading implies enlacing different signs and their meanings into a story that brings about people, places, and knowledge following a particular sense of history and relationship with the world. Past and present are weaved by the reading of the presence or signs that someone encounters. The encounter of signs as a way of creating narratives of futures and material for learnings can be thought through Kastuv Roy's pedagogical encounters.[19] Roy asserts that through the repetition "of collective beliefs, predispositions, prevalent wisdom, power relations, and existential imperatives, groups of signs become isolated and boundaries get drawn unifying them as this or that event."[20] This suggests that the signs are not simply there to be read but that the encounter and the reading of signs are part of a way of seeing that people learn in society. For example, in talking about "the school," a common isolation of signs would map buildings and outdoors. However, Kichwa stories might map a forest or a chakra—collectively cultivated land. Signs are human and more-than-human presences which people see and read as informed by history and their own experiences. How readers order signs allows them to tell stories that emerge within particular social relationships and histories. How people in a given context value a story makes it possible for that story to survive or not and thence to have or not have a place in the future. This is especially relevant to the returning of the past in the future that Kichwa thought proposes, wherein it is not a romantic idealization of a mythical past, but the reading of presences and the (re)construction of narratives for an otherwise future that draws on yet to be told possibilities and stories.

The re-reading of time-space from Kichwa stories' perspective unsettles the implicit narratives of progressing time and underdeveloped spaces that inform 'singularist' (hi)stories, which Ecuadorian government's "good living" exemplifies. The multiplicity of future(s) makes evident how official discourses about "good living" place groups of people and their epistemologies inside/outside society and its future(s). The multiplicity of future(s) in Kichwa stories, as the third section of this chapter shows, challenges the use of Indigenous language to justify extant developmental agendas and opens the door to question the role education has in making a good life and future and where the curriculum for such education is coming from. In the following sections, I present the constitutional and bureaucratic story of good living espoused by the government in the previous decade to then move in section four to the stories of good living as told by Kichwa storytellers.

Buen Vivir and Sumak Kawsay as a constitutional and bureaucratic story

Even though some scholars and political leaders have located the emergence of Sumak Kawsay (SK) concept or "good living" previous to the 2000s and as early as the late 1930s;[21] Sumak Kawsay as a story of good life for *all* and a guide for planning the future society of "good living" was not in use before the inclusion of "good living" in Ecuador's 2008 constitution.[22] To foster the "good living" as expressed in the constitution, the Planning and Development Secretariat of Ecuador, the Buen Vivir Secretariat, and the Ministry of Education produced documents that defined "good living." These three institutions storied Buen Vivir and Sumak Kawsay as a guide to the production of national and individual good futures. Official reports and websites have defined Buen Vivir thusly:

> Buen Vivir encompasses having free time for the contemplation and emancipation, and for the expansion and nourishment of the freedoms, opportunities, abilities and real potentialities of the individuals in a way that enables simultaneously the achievement of what is valued as a desirable life objective (materially as much as subjectively and without producing any kind of domination of each other) by society, territories, and diverse collective identities and for everyone—seen as both a universal and particular human being.[23]

Buen Vivir (BV) understood in this way follows Euro-American rationality of progress as built on "the assumption of slow, gradual, continuous change—cumulative, purposive, and self-driven"[24] present in international development notions. This definition of Buen Vivir is similar to what Amartya Sen describes as the capability approach to development and development as freedom.[25] The definition of "good living" as expanding "freedoms, opportunities, abilities" to achieve "what is considered valuable for life," is very close to Sen's proposal of "development as expansion of capabilities" or "substantive freedoms" for a person "to lead the kind of life that he or she has reason to value."[26] The kind of good living described by governmental institutions thence is ordered by international ideas of education that privilege freedom as individual "intelligent choice"[27] and imagines a world with under/developed populations and places that need opportunities and education to make better choices.

The definition of good living in education is also related to development, particularly to Jomtien's statements in the 1990 United Nations Educational and Scientific and Cultural Organization (UNESCO) which devised the Education for All (EFA) declaration. The definition of Buen Vivir offered by the Ministry of Education shows that relation clearly:

> Education and Buen Vivir interact in two ways. On the one hand, the right to education is an essential component of Buen Vivir because it

> allows the development of the human potentialities and in this sense guarantees equality in opportunities to all.
>
> On the other hand, Buen Vivir is a central axis of education in so far as the educative process must contemplate the preparation of future citizens with values and knowledge of how to develop the country.[28]

EFA described education goals in terms of "educational opportunities" for everyone to meet learning needs that include knowledge and values to "develop their full capacities, to live and work in dignity, and to participate fully in development."[29] Sumak Kawsay and Buen Vivir in education are conceptually closer to the United Nation Development Program (UNDP) and UNESCO reasonings of development goals than to the "wisdom of many ancestral groups of people, [i]n the Andean countries… the Kichwa people"[30] A starting point is to notice that development is not an ancestral objective but rather as mentioned earlier a product of the rationality of progress often traced to 19th century Europe stories about the trajectory of modern *Human* life.[31] That rationality, following Sylvia Wynter,[32] produces descriptive statements and stories about what it means to be human. That rationality universalizes development making it possible to talk of it as ever-present and forget its historical and cultural foundations. Importantly, this chapter points to stories showing visions and desires for living a good life and learnings to make that life possible, which escape the narrow definitions proposed by the government. Those visions, desires, and learnings can be formative of other futures as I will expand on in the next section. Thus, the bureaucratic logics of "good living" normalize national and international notions of development as to what is a good life and what is a good future. In so doing, these stories of "good living" close down the diverse possibilities of what "living well" and "good future" could mean.

As the guiding concept for constructing the society of good living, Buen Vivir became an objective of the secretariat for planning and development. As the pages of the national plan for Buen Vivir assert, "Buen Vivir must be planned, not improvised… If we know where we are going, we will arrive faster, because we will know how to avoid the obstacles that arise."[33] The secretariat designed a study "to guide public policy towards the execution of a project, a dream: the society of Buen Vivir."[34] This study involved the following: Projecting future desired scenarios and actions in the present to materialize those visions; monitoring evaluation systems to know the impacts of public management, and; generating timely alerts for decision-making.[35] The secretariat of planning and development turned "good living" into the product of a technocratic system to conduct life.

"Good living" as a bureaucratic story links education to the governments' vision for an ideal society based on technocracy and on international development theories, proposals, and objectives such as EFA. This way to think about "good living" in the future is unlike the Kichwa idea of a "good life" in the future that arises relationally, pregnant of multiplicities, and with unsettled aims.

Kichwa stories of future(s) and good lives

This section presents stories that friends, instructors and storytellers shared with me in conversations in Ecuador nearby Tena and in Quito. All the stories give a point of entry to Kichwa experiences of time, space and life wherein curriculum and notions of a "good life" in future(s) emerge. The stories point towards a kind of curriculum that is not necessarily planned but conjured. If we think of curriculum as knowledge that traces a path for being, then these stories talk about the encounters that conjure beings and their (hi)stories and lifestyles. It is part of a what Sylvia Wynter calls the "sociogenic" production of the world and who exists in it, that is narrating existence through an own sense of self and simultaneously disrupting stories that universalize it. These stories discuss diverse curriculum locales across schools, the land, elders and dreams simultaneously, which is important because they point towards ways in which Kichwa people learn about their future(s) and about making future(s) to live well.

The sun travels down river

Close to the Napo River students chatted. Someone asked a Kichwa instructor and storyteller named Pedro Andi from who had he learned so much about the forest. Pedro answered "from my step-father. He advised me to go home indiikuirinmi/when the sun hides inside the earth. He would wake us up very early because doce tuta pasajpi ña punllayauni/when midnight has passed the day is already coming into being. He was a great hatachik/waker. He would always wake us up chaupi tutamanda urayma/from midnight downriver." Pedro was talking about moments of the day using references to the physical movement of the sun. The day travels on the sky just as canoes do on the river. "Indil sikimanda llustimujpi hatarin"/from the base of the world the sun starts to stand up. The sunrise is something already happening when the sun goes down. In its trip, Pedro said, the sun goes inside the earth like the dead and it rises the next morning renewed. The sun hides, slides underground, and when it is time to come out, it travels from the place of midnight to the east, which is downriver. The day begins during the night and travels through the space of the past era (the underground) to transform in future(s) as they land in the day one experiences in the present. In the story that Pedro told, future(s) start in the present, go to the past, and unfold slowly when we cannot see them (underground, through the night when we are asleep).

The story presents two ways in which time relates to two spaces for the past that co-exist with the present. One is the underground, the world of the past era that emerges again at the end of this era. A Kichwa elder, Luisa Cadena, asserted that when this world ends the earth turns inside out, taking everything that we can see around us underground, and it will be the turn of the ancient people to live outside again. This story, which is common to Kichwa and Quechua people, speaks more to a horizon for the renovation of the world rather than to daily experiences of future(s).[36] However, it is important to have it in mind as a geographical reference of the past. The other space is the forest, which is home to the people from the past (trees, animals, spirits, dead, which are often overlapping and unstable categories) of this age who are most often awake during the night and asleep during the day. "When we die, we become part of the forest. Sometimes the plants eat from us. Sometimes we can turn into birds or pumas," Pedro said. "During the night the people from the past wake up, converse, and sometimes visit us in our dreams. For example, when we are sick the big old trees come as medical doctors and they heal us in our sleep." The past is always in interaction with the present and it actively affects future(s) because as future(s) come into being, they encounter the space and people of the past.

Future(s) arrive slowly through many signs. A rainstorm, for example, is announced by the color of the sky, the sound of the wind through the trees getting a bit stronger than usual, a dog at home getting anxious. All those are things that are part of the story of the storm coming into being and at the same time are signals weaving their presence for the farmer to learn that she will not be able to work that morning. Dreams have a similar way to utter signs and summon the presences of future(s) in that they are part of what is coming into being. For instance, Pedro said that dreaming about being bitten by an animal or a t-shirt getting stuck on a plant while walking inside the forest suggests caution and staying away from the forest that day. Similarly, dreaming of a car crashing announces a fight. Seeing the signals that come to us through the environment or through dreams changes how future(s) take place and what stories are narrated into being. Future(s) from this stand are the conjoining of stories which make some stories present for particular narrators, rather than the product of cause-effect dynamics that can be planned and foreseen for the good of all. The making of a particular future is a story told between the people of the past and the people of the present and everyone/everything around. Together they articulate both a sense of self and a sense of a world that has not yet taken place. In this sense, from a Kichwa perspective, the storytelling of the world is a product of the *homo narrata* that Wynter proposes and of other narrators, in so far as humans and other beings, have flexible boundaries between them and *people* has a larger than human meaning. That is, multiple beings collaboratively narrate into being a story of the world with possible relations for its beings (the human self among them).

Did you dream well?

A very common way of saying good morning among Amazonian Kichwa is: "did you dream well?" Good dreams and especially bad dreams are of great interest to most people because they are windows to see whom they will become and what will they experience. Interpreting dreams and talking about them is a very common practice as it is helpful to make the best of our days, avoid danger and be more open to receiving a future. There are certain dreams which have a straightforward interpretation. For example, if someone dreams that a molar (mama kiru) falls out, it means that the mother or father of that person will die. However, most dreams are decoded through a threefold process. Elodia Dagua, a Kichwa pottery master, explained that to interpret dreams people often meditate about the dream, then discuss it with family and friends and finally decide on its meaning. Deciding on the meaning is decoding a message that often comes from a person from the past (a dead relative, an animal or a tree). This process often happened while drinking guayusa during the early hours of the morning (3–5 am) when the sun has not come up yet. Nowadays that ritual is less common. Instead, dreams are usually interpreted during morning chores. Pedro explained that it is better to do the ritual very early in the morning "because we can hear better, there is no noise, only the forest sounds. The grandparents, the trees, and the animals from the forest are awake still." Interpreting future(s)'s signs was a thing to do at the moment when both the grandparents of the present and the grandparents of the past made their presence felt to give advice. This advice opens narratives and makes possible descriptive statements to story oneself into the day.

The guayusa ritual and the moments of interpreting dreams are spaces for the story telling of future(s). However more than predictions of what is coming ahead, these are spaces for collaborative uttering and contemplation of the signs already arriving from the space of the past to the present to narrate the story of the day. Knowledge is simultaneously emerging and being learned before storying the own self and day. Signs or presences are knowledge of whom we become; they are a kind of curriculum sprouting from dreams, conversations, and the environment. This kind of curriculum demands being attentive to what is coming together in order to become (continue to live) wisely. Education thence is not about bringing "the future" to the present to manage it. Instead, it is listening, interpreting, transforming, and storying a relational self.

These views imply that people are woven into future(s) as they see a story arising. Similar to how without a clock we can predict that the sun is about to rise when we see light growing in the sky and hear the songs of early birds, a story of future(s) can be uttered easily through signs that are arriving. Calculations of distant future(s), on the other hand, are unproductive in Kichwa thought because hundreds of actions, events, human and more-than-human people, all of which are simply changing every day, have not yet taken place and thus there are no clear presences yet to weave. The caution of trying to see too far ahead is due to the idea that one will come up with a stable delusional image that prevents readiness. In other words, beings, their life, and a good life come into being in collective stories.

Kichwa stories propose that in striving to actualize a plan the picture of what is coming together gets lost. Understanding the utterances of dreams and the presences of people from the past (ancestors and land) is important to narrate present(s) and future(s), the world and the people in it.

Memories of future(s)

In many Kichwa stories, dreamers have seen a story that already happened (in their dreams) but which is yet to take place. For example, if someone dreams about a family member visiting them, it is considered that that visit has happened at some level of existence and at the same time, the meaning of that visit speaks of a future event.[37] The dream is a real story that can be told differently once the person wakes up because there are presences during the night which might not be there during the day and vice versa.

When people interpret dreams, those dreams become a story which is telling events that effectively took place and foretell possible stories to be actualized when the person encounters the stories of others and of the land. Here I use "land" to translate the term "*sacha*," which in my understanding, is a living process that in different moments can mean forest, nature, chakra (cultivated land), and also a space marked by a sense of belonging and practices of dwelling with their cultural meanings. *Sacha*, or land as used in this paper, is a space that is constituted materially and culturally by (human and more-than-human) people, and which constitutes (human and more-than-human) people materially and culturally. *Sacha* has a more-than-human sociogenic quality as it is foundational for the articulation of a sense of self. In so far as the land plays an important role in who one becomes, the land can be thought of as a curriculum locale or a place from where the knowledge to narrate beings, lifestyles and futures emerges materially and in dreams.

In the dreaming and making of future(s), land as *sacha* has affinities to Ingold's[38] conception of a landscape; a living process "pregnant with the past"[39] that becomes part of us, just as we become part of it. Ingold's landscape is continually made and, at the same time, performed itself and together with people. The engagement of people with their environment physically, culturally, and spiritually forms the landscape, which similarly forms people. Landscape is the reiterative practices of the dwelling of past, present and possible future(s) generations. This is why Ingold says landscape is pregnant with the past. It carries all the actions of life that narrated (human and more-than-human) people and continue to narrate them as possible beings with particular lives and qualities. Land as sacha is an enduring life affected by and affecting other lives with which it has relationships.

Considering that signs in dreams come from material and virtual relationships to the land and (human and more-than-human) people, dreams can be understood as memories of past and future events and relationships. Mountains, forests, and bodies are a kind of archive in that they are constituted by memories in a certain order that continues their presence as what they are in their relationship with each other and their stories.[40] The land and people's relationships with each other allow for their survivance as a particular being. In this sense as well, a good/beautiful life in the future stems from those relationships. Tod D. Swamson and Jarrad Reddekop[41] have argued that a beautiful and mature life in Kichwa culture comes from becoming "physically connected to a network of plants, animals, rivers, and mountains… in their bodily and behavioral qualities."[42] Beauty is "a responsive adaptation to others… to survive and flourish."[43] Dreams, as memories of future(s) are an instrument for generating descriptive statements for the world and for people to narrate a beautiful story of existence. In this sense, anticipating a good life in future(s), does not only constitute a plan. Rather it is uttering changes in a series of relationships and articulating otherwise descriptions of the self and the world to re-tell a particular social and physical memory.

Relationships and encounters—which constitute the land, weave dreams and are part of what people become—leave traces. Those traces are constantly being (re)lived, are (re)emerging and are, in Kichwa stories, what shall be learned to survive and flourish. Surviving here has the quality of the stories that continue, like Gerald Vizenor's[44] survivance, "is an active sense of presence, the continuance of native [Indigenous] stories, not a mere reaction."[45] The stories that continue make the presence of people-and-land and are related—not reactionary—to national stories brought through national schooling, governance, and economic practices. All stories told and yet to be told intertwine in what people-and-the-land narrates, thus becoming the territory's histories. This challenges the singularizing and universalizing production of national history and frees the possibilities of storying the land-society.

In between landscapes

Pedro said he often takes the foam of the river and puts it on his skin to smell like the river. He says eating animal meat from the forest makes him more like the forest. Wild meat makes his body more beautiful and healthier. After spending a week in the Yasuni National Park I asked Pedro, would you like to go back and live in the forest? He said, "it would surely be nice, but it is not possible now to live in the forest. I have to work. In the forest and in the chakra (cultivated forest/land) life is good but hard." When we were in Yasuni Pedro was very happy to hear the sounds of the forest; he remembered he used to hear more animals when he was a kid. There used to be more fish in the water and more animals in the forest. In this and other conversations, he described the food of the city as food that makes the body sick. However, he talked about how he would want his descendants to study and work less hard than him, perhaps in an office. Even though the forest and the chakra might still be part of what he thinks makes him a mature, beautiful, healthy man able to have a good life; his idea of a good life was not necessarily to happen in the forest/chakra. His aspiration was that his grandchildren go to school in the town and later to a good university in the city.

This story shows how the experience of the city becomes part of the desires for the future. This is not to say that towns and cities are disconnected from the forest or the land. On the contrary, the systems of significations that produce both spaces have no fixed boundaries, particularly in terms of memories and stories. If we were to locate economy, labour, capitalist influences, colonialism in the city and an alternative to all that is in the forest we would be essentializing the life and history of entire ecosystems. Ideas and desires of good life and education and the future are encounters in between spaces which come about with no clear starting point. The in between is the space where stories thicken through relationships between people, community, environment, economy and more.

Walking paths, weaving curricula

Belgica Dagua was making mukawas/clay bowls when I asked her, "what do you think you need in order to have a good life?" "Yachaita/knowledge. You need to study and keep learning so that you have knowledge and you can have a good life." Belgica was then in her thirties and was a wonderful pottery master. She was about to graduate from high school and was looking for the economic means to go to university so that she and her son could have a more comfortable life. The desire for further education had to do both with her curious character and her awareness of the role of formal education in socio-economic mobility. Her view was, however, that young people were learning a lot at school but could learn more from the elders to know how to grow old and how to take care of themselves and others wisely.

Belgica's favourite stories spoke about "how the elders used to be, what they ate to get energy from the food, what dreams meant, and what dreams were useful for. When is the time when I have to plant." One day a student asked her, "what do young people need to learn?" She answered, "they need to learn how to walk in the forest. Today the younger people go into the forest toroc-toroc making noise with their boots disturbing the forest. They don't know how to walk." Belgica would wear long skirts and no shoes when walking into the forest. No matter how much it had rained, mud never reached her ankles. "I know what roots to step on," she said. She would hide between the vegetation and listen to the sounds of the forest. Walking in the forest is a way to learn because knowledge, as Belgica put it, should allow one to walk in any place and be received with joy by those present. Listening to the stories of elders and the forest allows for survivance. They give continuation to the presence of family/ancestors and place in what people learn about how to live. The survivance of stories maintains alive descriptive statements of who we are and what it means to have a good life that disrupt singularized and nationalized stories of good living. If we think of curriculum as a path that people walk in becoming a particular being with a particular lifestyle, these stories are curriculum and the forest one of their multiple curriculum locales with sociogenic power.

Encounters with the land at school

In the South of Quito, a small intercultural school proposes ishkay yachay or two "knowledges", which is a curriculum that relates all class subjects—mathematics, biology, social sciences, etc.—to Indigenous knowledge, particularly to learning with the land. According to Laura, one of the directors of the school, the most important part of education in the school is the cultivation of the chakra. The chakra at the school is a small piece of land where the kids cultivate potatoes and corn and let other plants grow among the crops that have medicinal use—like dandelion or llantén. Laura said that a chakra is not simply a vegetable garden because a chakra has a different order and purpose. It is a social and family space to share labour and grow your own grain. She considered the chakra to be the best teacher because a good life, she said, necessitates a healthy relationship with the land and with food. The chakra is as a story teller/teacher and the foundation for future(s) in that it is a space where knowledge and food is cultivated and passed on to sustain families, communities, and their possibilities of being. While walking by the chakra at the school she told me:

> The elders themselves sometimes say noooo, I don't know anything. Of course, because all their lives they have heard that they don't know. But then I tell them I'm sure you know something. For example, in my home, they taught me about the chakrita, how to plant, how to read the stars. Then they [the elders] say aaaah yes about that I do know, and they tell us [what they know]. We the Indígenas also need to decolonize ourselves and unlearn so that we can understand because we grow up with just one knowledge then we doubt what we know... Here in the school the mothers and the children also [learn together]. One kid says this plant in my community is called this name and I say aaah in my community that is called this other name and I ask them, what do you use it for at home? And the kid tells me, and other kids will know other uses and then I say oh very good let's write it down, so we don't forget.

For Laura the chakra allowed students to learn from the storytelling that emerges while being with the land or cultivating the land with family and community. Through the work in the chakra—moving the dirt, planting, harvesting—students learn from each other and parents and grandparents the names of different plants, how to use them for medicine, food, and crafts. Importantly, the chakra allows for a pedagogical encounter where land and community break together dominant significations of what is learning, who teaches, who is a student, and what is to know. The chakra emerges as a space for knowing what matters to make a good life in the future(s) that does not need to comply with the public education norm. Moreover, the chakra allows students to get involved in the production of food and medicine. This involvement prepares them to nourish their body, and to care for others by feeding them and treating illnesses with natural medicine. These learnings ensure survivance of ways of living well that are often overlooked as knowledge even by the same knowers/storytellers/elders/teachers. Decolonizing here, as Laura uses this word, has to do with making spaces for multiple ways of knowing and relating to the world. Particularly, the concern is to open spaces for Indigenous knowledge, life-stories, and ways of living of Indigenous families and ancestors to have presence in the future(s) of their own descendants and of the world more generally.

Final remarks: Dreamed educational landscapes

This chapter brings attention to stories weaved in narratives that map a reality oscillating between collective beliefs and predispositions common to the national order, and collective beliefs and predispositions that challenge that order. Buen Vivir and Sumak Kawsay entered Ecuadorian education as the instrument to achieve a dreamed society. However, that instrument, rather than opening the educational territory to multiple stories of good living, closed the future with goals and measurements for development. This chapter has focused on understanding how a "good life" in the futures might come about from a Kichwa perspective and what its relation to education might be.

This exploration allows questioning the claims that Sumak Kawsay and Buen Vivir, as used by the Ecuadorian government, are based on Kichwa knowledge, and to question the turning of "good life" into a "project, a dream"[46] to be measured and planned for all. To have an education and a national curriculum centered on the concern of students achieving a "good life" certainly sounds promising. However, the bureaucratized story of "good living" implies enforcing knowledge, education, behaviors, lifestyles, and ways of engaging with society. In other words, it implies the (re)production of a national norm, which is an average physical and moral to be met and which produces what is deviant in relation to it. Guarding the norm in terms of lives is a kind of anticipation of the future(s) to bring about a specific national future and in so doing singularizing the future(s). A normative planned good living for all legitimizes governmental interventions over the life and ways of knowing of people in the present. Closing down the multiplicity of stories of future(s) and "good living" through singularized and standardized conceptions, practices and curricula, comes at the expense of endangering the lives and knowledge(s) that do not fit the norm.

The stories in this chapter challenge the use of Kichwa language for coating extant developmental objectives. While not every Kichwa story might talk about "good living" in a similar way than that presented here, it remains necessary to pause and ask what are the consequences of striving for one kind of education and good life for all? What do people have to sacrifice to become differently and fit national expectations of progress? And why would it have to be so? Engaging seriously with the presences of land and (human and more-than-human) people, signs and dreams that abide in between the (hi)stories of progress, colonialism, the forest, the land, and other untold (hi)stories requires openness rather than a singularized future.

Simultaneously, the Kichwa stories presented offer a point of entrance to re-think curriculum as conjured, dreamed and collectively made through encounters. The pedagogical encounters that constantly happen between different (human and more-than-human) people as well as with the land are in themselves sociogenic spaces and point towards curricula locales that open up education as part of life paths and future(s). This implies a valourization of the collaboration among (human and more-than-human) people, perhaps from a chakra, a forest, a dream-interpretation gathering and other rich ecotones, in uttering and narrating stories about knowing and living through and beyond education systems. These are the kind of stories that free history that dismiss its own construction. They show the multiple descriptive statements possible for being a living being, for being people and for having a good life. The valorization of the encounters and collaborations as learnings, as creative spaces to make society and future(s) is a step in questioning the school as the primordial space for learning as well as what counts as learning. That inquiry would be an ally to the processes of survivance of Indigenous knowledge and a starting point to narrate future(s) otherwise.

Acknowledgements

Many thanks to all the Kichwa friends and teachers that made it possible for me to continue to learn Kichwa language and in doing so hear and better understand the stories presented in this chapter. Special thanks to Pedro Andi, Luisa Cadena, Belgica Dagua, and Elodia Dagua for their generous discussion of ideas and stories. Thank you as well to the Andean and Amazon Field School at Iyarina for facilitating the space for those encounters. I am always thankful to Armando Muyolema for our friendship and thought-provoking conversations. My gratitude to Matt Galway for all his writing advice and support, and to all the reviewers whose comments helped me strengthen this chapter.

Notes

1 Kichwa is a language and culture present in what is now Ecuador and part of the greater Quechua family that extends with several local variations throughout the Andes. See more on its history and geography in Cerrón-Palomino, Rodolfo. *Lingüística Quechua.* Cuzco, Perú: Centro de Estudios Rurales Andinos "Bartolomé de las Casas", 1987.

2 Martina Faller and Mario Cuellar, *Metáforas del tiempo Quechua [Metaphores of Quechua Time].* (Nijmegen, Instituto Max Planck de Psicolinguistica, n/d). https://personalpages.manchester.ac.uk/staff/martina.t.faller/documents/Faller-Cuellar.pdf

3 There is a decade of discussions about Sumak Kawsay and Buen Vivir, its meaning, impacts and the politics of translation. Authors like Alberto Acosta ("El Buen Vivir como alternativa al desarrollo. Algunas reflexiones económicas y no tan económicas," *Política y Sociedad* 52, no2 (2015): 299–330.) have argued that Sumak Kawsay lost its meaning when translated into Spanish as Buen Vivir, other authors like Rafael Dominguez, Sara Caria, and Mauricio Leon, ("Buen Vivir: Praise, instrumentalization, and reproductive pathways of good living in Ecuador," *Latin American and Caribbean Ethnic Studies*, 12 (3): 133–154, 2017) instead argue that both Sumak Kawsay and Buen Vivir are an invented tradition. In this chapter I am not interested in tracing the origins of Sumak Kawsay or Buen Vivir and I do not

assume that one discourse has an Indigenous origin or legitimacy and the other does not. Nor do I assume Indigenous politics and views to be homogeneous or monolithic across different groups. I propose instead that all the stories available are that, stories, that different groups of people weave into their way of thinking, advocating and desiring a particular kind of good life. For this reason, I use "good living" to address any story that talks about that intention from different perspectives. Part of the argument of this text is however, that the bureaucratic story of "good living" is rather dangerous and violent to other views and desires, particularly those of certain Indigenous groups, for what might be called a "good life".

4 See Secretaría del Buen Vivir. Buen Vivir, Sumak Kawsay ¿Qué? ¿por qué? ¿para qué? y ¿cómo? Mayo, 2015. http://www.secretariabuenvivir.gob.ec/wp-content/uploads/2016/07/Qué-Por-qué-Cómo-libro.pdf

5 Ben Anderson, "Pre-emption, precaution, preparedness: Anticipatory action and future geographies." *Progress in Human Geography*. Vol 34 (6): 777–798, 2010.

6 Katherine McKittrick, ed. Sylvia Wynter, "Unparalleled catastrophe for our species? Or, to give humanness a different future: Conversations," *On Being Human as Praxis*. Durham: Duke University Press: 9–89, 2015.

7 Silvia Rivera Cusicanqui describes ch'ixi as "the parallel coexistence of multiple cultural differences that do not extinguish but instead antagonize and complement each other. Each one reproduces itself from the depths of the past and relates to others in a contentious way." See Silvia Rivera Cusicanqui. "Ch'ixinakax utxiwa: A reflection on the practices and discourses of decolonization." *The South Atlantic Quarterly* 111 (1) (Winter 2012): 105.

8 I borrow this term from Boaventura de Souza Santos. In his book Epistemologies of the South: Justice against Epistemicide. (London: Routledge/Taylor&Francis), 2017, Santos asserts that European critical traditions have drawn a line that produces difference and in terms of knowledge produces science, philosophy, theology on its side of the line, and incomprehensible, magical, idolatrous thinking on the other side of the line. That division discards and invisibilizes knowledge that escapes Euro-American thought and history. Bridging that abyss is fostering an ecology of knowledges that can bring new stories and terms to our understandings of social phenomena.

9 Wynter, *On Being Human*, 25.

10 Sylvia Wynter, "Towards the sociogenic principle: Fanon, identity, the puzzle of conscious experience and what it is like to be 'black.'" *National Identities and Sociopolitical Changes in Latin America*. Eds Mercedes Duran-Cogan and Antonio Gomex-Moriana. New York: Routledge: 34, 2001.

11 See Sylvia Wynter, "Unsettling the coloniality of being/power/truth/freedom: Towards the human, after man, its overrepresentation—an argument," *The New Centennial Review*. 3 (3): 257–337, 2003.

12 Gerald Vizenor, *Manifest Manners*. Hanover: Wesleyan University Press: 11, 2014.

13 Ibid, 11.

14 Eduardo Kohn, *How Forests Think: Toward an Anthropology Beyond the Human*. Berkley, Los Angeles, London: University of California Press, 2013.

15 Ibid, 217.

16 Ibid, 76.

17 Ibid, 216.

18 Ibid, 212–213.

19 Kastuve Roy, *Teachers in nomadic spaces: Deleuze and curriculum*. New York: Peter Lang:,1–18, 2013.

20 Ibid, 14.

21 Jose Inuca, "Genealogia de alli kawsay/sumak kawsay (vida buena/ vida hermosa) de las organizaciones kichwas del Ecuador desde mediados del siglo XX [Genealogy of allí kawsay/sumak kawsay (a good/beautiful life) one of the Kichwa organizations

of Ecuador since the middle of the XX century]" *Latin American and Caribbean Ethnic Studies*, Vol, 12 (3): 155–176, 2017; Carlos Viteri, "Visión indígena del desarrollo en la Amazonía [Indigneous view of development in the Amazonia]" *Polis*: 3, 2000). https://journals.openedition.org/polis/7678#quotation

22 This is an important shift considering that the creation of independent nations in the continent gave continuation to colonial relations and initially did not recognize Indigenous peoples as their citizens. The inclusion of Sumak Kawsay in the constitution was a ground-breaking event in terms of the recognition of Indigenous knowledge. Regrettably the government later co-opted Sumak Kawsay and instrumentalized it for its own developmentalist and republican project. See Aníbal Quijano, "Bien vivir": entre el "desarrollo" y la des/colonialidad del poder, *Cuestiones y horizontes: de la dependencia histórico-estructural a la colonialidad/descolonialidad del poder*. (Buenos Aires: CLACSO, 2014); Boaventura de Souza Santos, "El socialismo del Buen Vivir," *America Latina en Movimiento*. No. 452, (February 2010): 4–7; Acosta, Alberto. "El Buen Vivir Como Alternativa Al Desarrollo. Algunas Reflexiones Económicas y No Tan Económicas." *Política y Sociedad*, vol. 52, (2), July 2015, p. 299–330.

23 Secretaría Nacional de Planificación y Desarrollo – Senplades, *Plan Nacional de Desarrollo Plan Nacional del Buen Vivir 2013–2017 [National Plan of Development National Plan of Good Living]*. Ecuador: Senplades. Recuperado del URL http://www.buenvivir.gob.ec/versiones-plan-nacional

24 Robert Nisbet, *History of the idea of progress*. New Jersey: Transaction Publishers, 2009.

25 Amyrta Sen, "Development as capability expansion." In Fukuda-Parr, S. et al, *Readings in Human Development*. New Delhi and New York: Oxford University Press, 2003.

26 Ibid, 87.

27 Ibid, 55.

28 Ministerio de Educación del Ecuador, *Curriculum de educación básica Ecuador*. Quito: Ministerio de Educación del Ecuador, 2016. Recuperado del URL https://educacion.gob.ec/wp-content/uploads/downloads/2016/08/Curriculov2.pdf

29 World Conference on Education for All: Meeting Basic Learning Needs, *World declaration on education for all and framework for action to meet basic learning needs adopted by the World Conference on Education for All: Meeting Basic Learning Needs, Jomtien, Thailand, 5–9 March 1990*. New York: Inter-Agency Commission (UNDP, UNESCO, UNICEF, World Bank) for the World Conference on Education for All.

30 Secretaría del Buen Vivir. "Saberes ancestrales: lo que se sabe y se siente desde siempre," accessed 2016. http://www.secretariabuenvivir.gob.ec/saberes-ancestrales-lo-que-se-sabe-y-se-siente-desde-siempre/

31 See Robert Nisbet, *History of the Idea of Progress*. (New Jersey: Transaction Publishers, 2009); Bruno Latour, *We Have Never Been Modern*, trans. Catherine Porter (Cambridge: Harvard University Press, 1993) and Sylvia Wynter, "Unsettling the coloniality of being/power/truth/freedom: Towards the human, after man, its overrepresentation—an argument," *The New Centennial Review*. 3, (3): 257–337, 2003.

32 See Sylvia Wynter, "Unsettling the coloniality of being/power/truth/freedom: Towards the human, after man, its overrepresentation—An Argument," *The New Centennial Review*. 3 (3): 257–337, 2003.

33 Secretaría Nacional de Planificación y Desarrollo – Senplades, *Plan Nacional de Desarrollo Plan Nacional del Buen Vivir 2013–2017*. Ecuador: Senplades, 14, 2013. http://www.buenvivir.gob.ec/versiones-plan-nacional

34 Ibid, 62.

35 Ibid, 18, 36, 62.

36 Armando Muyolema, personal communication with the author, June 25, 2017.
37 Ibid.
38 Tim Ingold, "The temporality of landscape". *World Archaeology*, Vol 25 (2): 152–174, 1993.
39 Ibid, 153.
40 Armando Muyolema, personal communication with the author, June 25, 2017.
41 Tom Swamson and Jarrad Reddekop, "Looking like the land: Beauty and aesthetics in Amazonian Quichua philosophy and practice." *Journal of the American Academy of Religion*, Vol.85 (3): 682–708, 2017.
42 Ibid, 685.
43 Ibid, 703.
44 Gerald Vizenor, *Survivance: Narratives of Native presence*. Lincoln: University of Nebraska Press, 2008.
45 Ibid, 19.
46 Senplades, *Plan Nacional BV*, 62.

References

Anderson, Ben. "Pre-emption, Precaution, Preparedness: Anticipatory Action and Future Geographies." *Progress in Human Geography*, 34 (6): 777–798, 2010.

Cusicanqui, Silvia Rivera. "Ch'ixinakax utxiwa: A reflection on the practices and discourses of decolonization." *The South Atlantic Quarterly*, 111 (1): 105, 2012.

de Souza Santos, Boaventura. *Epistemologies of the South: Justice against Epistemicide*. London: Routledge/Taylor & Francis, 2017.

Dominguez, Rafel, Sara Caria, and Mauricio Leon. "Buen Vivir: Praise, instrumentalization, and reproductive pathways of good living in Ecuador." *Latin American and Caribbean Ethnic Studies*, 12 (3): 133–154, 2017.

Faller, Martina and Cuellar, Mario. Metaforas del tiempo Quechua [Metaphors of Quechua time]. Nijmegen: Instituto Max Planck de Psicolinguistica. https://personalpages.manchester.ac.uk/staff/martina.t.faller/documents/Faller-Cuellar.pdf (n/d).

Ingold, Tim. "The temporality of landscape". *World Archaeology*, 25 (2): 152–174, 1993.

Inuca, Jose. "Genealogia de alli kawsay/sumak kawsay (vida buena/vida hermosa) de las organizaciones Kichwas del Ecuador desde mediados del siglo XX [Genealogy of alli kawsay/sumak kawsay (good life/beautiful life) of the mid XX century Kichwa organizations of Ecuador." *Latin American and Caribbean Ethnic Studies*, Vol. 12 (3): 155–176, 2017.

Kohn, Eduardo. *How Forests Think: Toward an Anthropology Beyond the Human*. Berkley, Los Angeles, London: University of California Press, 2013.

Manga Quispe, Eusebio. "Dos concepciones espacio-temporales para dos mundos. Ñawpa y ñawpa-n: encaminadores de kay pacha [Two time-space conceptions for two worlds. Ñawpa y ñawpa-n: pathways of the kay pacha]." *Ciberayllu*. http://www.ciberayllu.org/Ensayos/EMQ_Concepciones.html, 2010.

Ministerio de Educación del Ecuador. "Curriculum de educación básica Ecuador: Ministerio de Educación del Ecuador", 2016. Recuperado del URL https://educacion.gob.ec/wp-content/uploads/downloads/2016/08/Curriculov2.pdf.

Muyolema, Armando. (2017, Junio, 25). Personal interview.

Nisbet, Robert. *History of the Idea of Progress*. New Jersey: Transaction Publishers, 2009.

Potosi, Fabian. (2017, July 25). Personal interview.

Roy, Kaustuv. *Teachers in Nomadic Spaces: Deleuze and Curriculum*. New York: Peter Lang, 1–18, 2003.

Secretaría Nacional de Planificación y Desarrollo – Senplades, "Plan Nacional de Desarrollo Plan Nacional del Buen Vivir 2013–2017" [National Development Plan of Good Living 2013–2017]. Ecuador: Senplades. Recuperado del URL http://www.buenvivir.gob.ec/versiones-plan-nacional.

Secretaría del Buen Vivir. "Saberes ancestrales: lo que se sabe y se siente desde siempre [ancestral knowledge: what we have always known and felt]." 2016. http://www.secretariabuenvivir.gob.ec/saberes-ancestrales-lo-que-se-sabe-y-se-siente-desde-siempre/.

Sen, Amyrta. "Development as capability expansion." In Fukuda-Parr, S., et al *Readings in Human Development*. New Delhi and New York: Oxford University Press, 2003.

Swamson, Todd. (June, 2017). Personal communication.

Swamson, Todd and Reddekop, Jarrad. "Looking like the land: Beauty and aesthetics in Amazonian Quichua philosophy and practice." *Journal of the American Academy of Religion*, 85 (3): 682–708, 2017.

Vizenor, Gerald R. *Survivance: Narratives of Native Presence*. Lincoln: University of Nebraska Press, 2008.

Vizenor, Gerald. *Manifest Manners*. Hanover: Wesleyan University Press, 2014.

Wynter, Sylvia. "Towards the sociogenic principle: Fanon, identity, the puzzle of conscious experience and what it is like to be 'black.'" In *National Identities and Sociopolitical Changes in Latin America*. Mercedes Duran-Cogan and Antonio Gomex-Moriana, eds. New York: Routledge, 2001.

Wynter, Sylvia. "Unsettling the coloniality of being/power/truth/freedom: Towards the human, after man, its overrepresentation—an argument." *The New Centennial Review*, 3 (3): 257–337, 2003.

Wynter, Sylvia. "Unparalleled catastrophe for our species? Or, to give humanness a different future: Conversations." In *On Being Human as Praxis*. Katherine McKittrick, ed. Durham: Duke University Press, 2015.

3 Chi uwach loq'alaj q'ij saq

The sacred existing in knowing/learning from space/time

María Jacinta Xón Riquiac

Introduction

Dialoguing about time/space being Maya-k'iche' in contemporaneity is a narrative challenge of content. It is a responsibility of objectivity in regard to a stigmatized epistemology as a set of magical, mystical, subjective, folkloric, colorful, and wild beliefs. The civilizational process through Christianization and racialization have been policies from Spanish colonization from the first centuries of the conquest to the current state in the modern era. What the world knows about the conceptualization of time/space of Indigenous peoples in the Americas and the world are visions and fictions constructed from ethnology, anthropology, and archaeology and all take conceptually the prefix "etno." The fact that this chapter is narrated and explained making use of a colonial language (Spanish) is a paradox of our current reality. In fact, the representations made by Indigenous peoples like us, who have gone through civilizational processes that give "authority" to represent us before the academy to which we aspire to belong, are already intersected by a racialized vision of ourselves.

Racializing is the process through which a person, in this case an Indigenous person, goes through an ideologic formation and training of systematic denial of being "Indigenous". This process is undertaken strategically from various state apparati in complicity with global policies of industrialization and homogenization to subordinate Indigenous people for the benefit of consumerism. Is it possible to conceptually define or penally designate as "racist" an act, for example, when an Indigenous person pejoratively calls another person "indian" to show certain civilized superiority over them when the former considers themselves "less indian" given their educational level, their political allegiances, and creed? The state and its assimilationist policies have been responsible for this manifestation of internalized racism for centuries, which in this chapter I call a *process of racialization*. Much of this process has been undertaken by Guatemala's educational system up until the first years of the 21st century. Schools forbade and punished the use of Mayan languages and distinctive Mayan clothes in schools. Other processes that have contributed to racialization include the establishment of Catholicism, and the proliferation of Pentecostal churches. These, among others, have been key in the enactment of larger national policies to make "less Indigenous" or "not so Indigenous" Indigenous people.

Writing about time/space of the K'iche People in this chapter is an aspiration to write a narrative that we would share; meanwhile some of us are able to, in one way or another, continue the science of our grandmothers and grandfathers and their grandmothers and grandfathers as we hand-shell the corn, still native despite public policies that attempt to convince us of the low economic profitability of cultivating it, while we *nixtamalizamos*[1] the corn with *cal* or ashes; while we make corn into *tortillas*.[2] Hegemonic feminism[3] defends, intentionally or not, capitalist precepts when it affirms that the domestic space is an *a priori* and only an oppressive space. If indeed patriarchy defined the domestic space as *exclusively for women*, it does not mean interment lethargy for women awaiting to carry out domestic duties as thoughtless automatons. The domestic space has also been a creative and experimental one in the production of knowledge and resistance for and by women. Somehow feminism omits that reproducing life through cultivating and harvesting seeds in the fields or in a family garden is indeed the biggest act of rebellion possible against the capitalist system. Cultivating and making the elements we cultivate into food for our collective survival is an admirable revolutionary act in the era of Monsanto's food colonization. For instance, what we discuss during the time we prepare and eat our food near the fire reminds us of the origin of the universe, of the world, the life of our absent and transcendent mother who told us that without fire no warmth can exist at home. A revolutionary act is enacted by the fire as we discuss about our past, the peoples of now and the dreams we dreamed at dusk, while the youth share the world news and memes on Facebook. Ultimately being Indigenous is not synonymous with cultural or technological death, neither are traditions synonymous with being "on the verge of extinction."

Uj, ri qawinaq qib'/We, the peoples here

Despite the fact that dreams are complex and confusing, Indigenous peoples in this corner of the globe believe that they are omens, warnings, visions of the past and the future, a means of communication belonging to those who no longer have a voice to our ears. Therefore, we tell those dreams, we analyze them, we consult with them and expect an outcome. It is like when we hear the omens of the *Tukur*/owls or *Piscoy*/Cuco squirrel.

Dreams do not seem to have limitations of logic. When we dream, all that is impossible seems possible. What is possible is what we aren't already expecting. When dreams are described, logic is not a condition either. So, that day, many years ago I had a dream. I was told: Three dreams, three people that dream the same dream… 12 years later I dreamed the outcome of that augury.

> A woman sleeps. It is already September. It is cold and finally there is rain! This year the rain did not arrive when the *torb'al jab'*/migrating birds went through the skies. There were church services, evangelical services, and many

others bought candles, flowers, *pom/copal* (local incense) to ask for the life that only rain can give. The fruits of the earth did not grow or die in their attempt to survive. ¡*Xajalataj wa' uwuch ulew*!/The times changed! The Elder said. ¡*Qariq numik wa'*…!/We will be hungry…! She heard the morning before.

It is a sunny day. "It feels like a November morning," she tells herself. She knows she is asleep and does not remember she had woken up. Each ray of sun that touches her body feels warm and she does not doubt it is real. Well, at least she wants to believe it is real. She looks at the horizon and sees two simultaneous dreams, two simultaneous lives. But there, the horizon does not exist. She is already looking at the two women. Far, very far… they cannot see her. Ever so near that she can see once and again her stories from before being born – her fears, her faith, what she wanted from the after, from one day.

> I can recognize them from some place, from some memory, from some tale my *wati't numam*/grandparents used to tell once in a while. Each story/world seems to go by separately in each of my eyes. The woman in my right eye is carrying flowers in her hand. And the woman in my left eye has a son called Mash, and two nephews called Erick and el Nanas.

She knows it is a dream. The sun is warm, like a mother's embrace given to one she has not seen in many years. It is really comfortable being there. The clouds take shape and dissipate in the sky, as if in a rush to leave or arrive. *Pech Mam*/grandfather kept her company, whispered things to her, showed her aspects of stories perhaps of the past, perhaps of the future, like when she was a little girl and she woke up afraid.

> Why do you always have flowers in your hands? Now I remember. I saw when the midwife told your mother that a man would find you when he saw the flowers… but, do you have to carry flowers everywhere? All the time? It is as if it was the only sense in your life. Carry flowers… No! Many years have passed. You were just a girl when they told you that story. Did you misunderstand it? Well, I think so… It is not a husband what she predicted for you. I remember that story… *Ri ajpatan achi, kuriq rib' ruk' echi'rek'am kotz'i'j*/carrying flowers that man will find you to fulfil a mission. She also said, *Xuquje' pa ri jun q'ij ri' jun nimalaj xik karapan chi jumul*/that same day a giant eagle will fly again.

In another place, perhaps another time… I am sure indeed that both moments were happening simultaneously even though she was looking through just the left eye. Erick's and Nanas' mother, Rebecca needed to cross the bridge to pay a visit to the other side. There was a bridge that joined the two sides; it was a narrow bridge for meeting and crossing. It seemed as if two gigantic birds had joined the tips of their wings creating a narrow crossing. Indeed, in the story she had managed to know of the whispers by the *ujer winaq/*

of the peoples of before now. They said that both places were two fossilized eagles. "It cannot be!" Mash also wanted to go to the other side. Mash's mother's heart felt an asphyxiating sense of despair. Mash would become distant from his mother while the world they knew would crumble into a thousand pieces.

> Listen to me please! Don't let them go! Mash's mother. There will be an earthquake and they will all die on the bridge. If they turn back now, they will be able to make it before the earthquake begins... You will die of loneliness if something happens! Imagine if they were to get trapped, if you cannot help them, if you cannot see them again? Don't let them go. Or better... go with them. Don't leave them alone...
>
> What strange dreams I've been having for so many days... Kak'is nuk'ux/my souls sadden with anguish." Mash's mother wonders. "Would it just be my thoughts...? Mash is just with his aunt Rebecca. I feel as if the deep roots in my heart were being pulled.
>
> Mash's mother. No! No, they are not premonitions! Listen to me please, go find them, don't separate yourself from them.

She knows that the earth will tremble, that perhaps no place is safe. She also has a son and she know of no bigger anguish than not being near him, not knowing of his *se'j xumal*/fruit in flower. She was concerned about her Mash... It would ease her to know that perhaps she would be dreaming, far from there, but very close to his heart. Overwhelmed, Mash's mother ran to find them. She arrived a few minutes before they crossed the path. She felt her own heartbeat again when she found her kids. Oh no! Suddenly she remembered that the earth would wake up from its long dream—because the earth also sleeps and dreams. "Take them!" She told Mash's mother. "We have to go far, as far as the mountains," Mash's mother shouted as loud as possible to alert Rebecca, Mash, and Nanas.

And the women with the flowers? She remembered again. She was there. Also full of anguish, letting the days, months and years pass by. She awaited the prophesy. She waited for the love of a man. She blamed herself whenever she forgot the flowers. "What if it was me who misinterpreted the prophesy?" She wondered.

> Hey, listen. Yes, you, the one with the silly flowers... How can you let your life pass by waiting for a man that you believe will love you because you carry flowers? I can see what happened before you were born. I heard what the midwife told your mother. I can also see what will happen to you. Listen to me. Forget the flowers. Perhaps the midwife was confused. Or perhaps she wanted to say something else. No! You do not want to listen to me. You don't even believe they are premonitions.

Oh no! The two worlds have begun to crumble, the one from the left eye and the one from the right eye. No! The woman with the flowers stays undefeated before the earthquake while she waits. Destiny is what she calls it... How irrational and absurd!

No! It cannot be! There is a man. He appears to be fully focused trying to connect two cables and press two buttons. It seems as if he is trying to control a machine from a control tower. His hands call her attention. His hands touch each centimetre of the place. It is a miniscule and decaying place, full of cables and buttons. She is surprised as each cable is perfectly ordered in the direction of a special button. But they are also interconnected. They look like a perfect woven pattern. Something in that man calls her attention. His hands work in perfect synchrony. He knows every space filled in that place. What? How? He is blind!

> Woman who carries the flowers, the man you wait for…is…blind! Your time waiting is futile. It is not a love. It is not a husband. Do you understand he will never see the flowers you carry? You have been carrying flowers uselessly all your life. I told you so. Your mother misunderstood the prophesy. The world is crumbling down. Destiny does not exist. What I saw and heard is not your destiny either. Why would someone want to know your life before you live it? While waiting you would never know that the man was blind! What am I saying to you? How would I know that the man who is blind is the man of the prophesy?

"Destiny does not exist!" She concluded. "But, how did I know the earthquake would happen? I saw him. He, the one who has been travelling for a while to the now and the after now. *Pech Mam*/grandfather Pech – He is the one that showed it. He? He saw the wrong thing? Perhaps the woman with the flowers knows that she defies probability, that she defies the impossible? That's always what believing in destiny means, no? No, she doesn't believe so, but she cannot not believe so." She told herself. But she heard that in other places people see the future. She knows what her eyes did not see but which was seen by other eyes, someone else's eyes, the eyes of those who dream, perhaps the eyes of the midwife.

The earthquake began. Two paws slowly appear on the sea. Chucks of earth fall off between blurry angles and murky waters. At the same moment, a glowing root takes shape in the sky and shocks a man. The impact electrocutes him. His eyes closed open by force of habit. But he cannot see with the eyes he knows. He cannot see the iris. Only sclerotic remains in his memory. Everything happens so quickly. She knows that she is quickly waking up, as quickly as the *kaqulja*/lightning that has just struck.

Before she wakes up, she still has some time to tell her:

> Woman with flowers, after all perhaps it still will be unknown what one is to live before having lived it. Who knows, perhaps destiny cannot be really known. Perhaps the man will see your flowers. Perhaps there are other ways of seeing.
>
> *Oxib iq'*/Day three of the Moon/Wind/Jade – 8 October 2019

This dream could be read as an illogical narration of a surrealist impossible, but here it refers to the *Aj q'ijab'*/counters of days. The *Tz'ite'*, as I will explain later in the chapter, are an important example to get closer to the comprehension of how millenary and contemporaneous Indigenous[4] peoples conceive time-space, omens, anguish, hope and the future.

Ujertaq winaq/The people of before/The people of always

Being born and growing up in a Maya K'iche' family where the maternal side refused to become Christian for centuries is perhaps one of the reasons why I am so attentive to what one dreams. Like many others in this part of the world, every day we hear stories of how our grandmothers and great grandfathers, great aunts, mothers, fathers, etc., dreamed omens with warnings that became avoided or avoidable realities. Following conceivable logics by modern rationality, my ancestors resolved future enigmas that they could or could not avoid. Our quotidianities are full of narratives, of dreams: What day was dreamed before midnight or at dawn? Those who listen to the dream always have an opinion about the symbolisms within the dream. They share their opinion in regard to what dreams predict in relation to their own dreaming experiences. We believe that somehow the veil of modesty, the visual limitations, the limitations of knowledge, movement, and actions imposed by the first men[5] are overcome while one dreams.

Adrián Inés Chávez[6] in his book *Pop Wuj. Poema Mito Histórico Kí-ché*, translates the K'iche' manuscript "transcribed by" Fray Francisco Ximenez from 1701 to 1703. The book includes the following paragraph:

> They were said to be Constructed, Formed; they had no mother or father, only them we named because they were not born from a woman but were created by the Architect, Maker, Created, Created Male. It was only a miracle, in a myth it was how the construction and formation by the Architect, Maker, Created, Male Created, Tepu, Hidden, Serpent, they were reproduced as peoples; they made themselves peoples, they saw, heard, walked, touched; they were good peoples; chosen faces, those akin, they could bread; they could see, indeed their vision was far reaching, they could see a lot, they knew a lot, everything that was under the sky. If they saw, at the moment they would observe, they examined everything of the sky and the earth. There was no obstacle to see everything. They did not have to walk first when they wanted to see what was under the sky. They were in the same place when they looked. Much was the wisdom they had; their vision could go through trees, stones, lakes, seas, mountains, and coasts. In truth they were sacred peoples Blom Kitze, Blom akab, Maj Ukutaj, Ik Blom.[7]

They were at once asked by the Architect, the Form-giver:

> "How do you feel your existence? Do you see? Do you hear? Is our speech good? Is this your walk? Look now, see what is below the sky. Can you see the mountains and coasts? Make the effort." They were told.

Then they were done seeing everything under the sky. They thanked the Architect, the Form-giver:

"Truly, twice I thank you, three times I thank you for having made us people, we settled, we reproduced, we talked, we heard, we meditated, we moved, we felt, we knew far from close; Did not we recognized big and small, heaven and earth? Thank you very much, we became people, we are built, we were formed by you our grandmother, by you our grandfather," they said.

At once they appreciated their construction, their formation; they finished knowing it all: cusp of heaven, sides of heaven; the interior of heaven and earth. But this did not seem right to the Architect, the Form-giver.

"It's not good what the people we built, the people we formed, have said. Because they said they knew everything, big and small," so they said.

They thought again, the Created, Created Male:

"What are we going to do with them now so that they look closely? What should we do for them to look over a small surface of earth? Because what they say is not good. Are not they simply built-up people? Have not we formed their names? Will they exist first as gods? Will it be worse if they don't multiply and fail to increase as they speak and by dawn? What if there is no way for them to reproduce at all? We could just break them down, so they become a few desires. Because it is not good, that is what we feel. Are they going to match our deeds? Will their wisdom go far away, seeing it all? The Spirit of Heaven, One Foot, Last Lightning, True Lightning, Infinite, Hidden Snake, Created, Created Male, Ixpiyakok, Ixmukané, Architect, Form-giver, told each other.

They tried to amend their constructions and formations, for which Spirit of Heaven only tarnished their eyes, they were somewhat blind, as if Spirit Heaven had cast vapor over a mirror; Spirit Heaven blinded their eyes, they could now only closely look, they could only see where they were.

This is how they lost wisdom and all mastery the first four men, and so was the construction and formation of our first grandparents, first fathers by the Spirit of Heaven, Spirit of the Earth.[8]

Loq'alaj k'aslemal/Exist

For contemporary Mayan K'iche', there are still several ways to exist in space/time. To exist in the *Chuwuch q'ij saq*/the existence of time-space. The first is the *Q'ij*/day, which is measured in the *Nik'arajik*/major ripples – or major waves of time – which have a frequency of 12 hours. The major ripples are midday and midnight. The *Tztolajinik*/rounds, or minor ripples, also have a frequency of 12 hours. They ondulate approximately at 6 a.m. and at 6 p.m. More than a linear conception of time, it is a cyclic and dynamic conception unfolding like a spiral. For example, the rotation of the Earth and the translation of the Earth around the sun, are unquestionable. But, in this conception, even though the lapses of a day (clarity and darkness) and the lapses of the solar year are the same, these happen every time in different space/time. Yesterday lasted the same as today and will last the same time tomorrow, but they are all different days. Just like the movement of the Earth around the sun and the movement of Venus around the sun, and so on, are similar but not the same. The agricultural year in the Western area of the land that is now called Guatemala is measured in the *Q'alaj*/rainy season and the *Saq'ij*/dry season.

The image of a snake and its movement represent very well the undulating movements of the present as it exists spiraling towards the future and towards the past. Perhaps it is for this reason that the snake is one of the most recurrent representations of the passage of time since the pre-Hispanic era. During the day, at 12:00 p.m. or midday, and 00:00 through the night, feelings and ailments are more profuse. In the moments of major undulations, the sick gets worse, fever increases and suffocates those who suffer from it. In the moments of minor undulations, it is believed that people should not be in certain places, such as under the leaks of roofs, because the collectors of *Jaleb'*/souls or spirits can get confused and take away the souls or spirits of the people in those places.

A time-moment-place important mainly for pregnant women is the counting of *Ik'alal*/moons, because these will mark women in the months of pregnancy. Observing the moon is a way to prepare while waiting for the moment to give birth. *Ri iyomab', Ajq'ijab', Ri ajchapalbaq*/midwives, spiritual guides and Indigenous traumatologists, grouped together in the *Asociación Médicos Descalzos* (Barefoot Medical Association)[9], concluded in their 20 years of monthly meetings that knowledge of the Sacred Mayan Calendar, *Cholq'ij* in K'iche' language, structures individual and collective life in Mayan communities through the knowledge of *Q'ij alixik*/day of birth. "*Cholq'ij* is a time counting system, structured in thirteen groups of twenty days",[10] *Batz', E, Aj, Ix, Tz'ikin, Ajmaq, No'j, Tijax, Kawoq, Ajpu, Imox, Iq', Aq'ab'al, K'at, Kan, Keme, Kiej, Q'anil, Toj and Tz'i'*.[11]

> The progression of the days is obtained from the articulation of a 20-day cycle, *Nawalib'*,[12] with one of the 13 energies' cycle. Therefore, there are 260 progressive combinations between the two cycles and, each day is designated by the specific relationship of one of the 20 *Nawalib'*, with one of the 13 energies.[13]

In Indigenous communities today, the *Q'ij alixik*/day of birth is still observed, specifically among those who still practice the ancient rituals. A lot of attention is paid to the 260 combinations of the 13 groups of 20. This is because it is considered that there is a tendency of action, reaction, relationships with dreams, relationship with the *Ujer winaq*/the past people and with the gift for collective service one should develop, etc. All those tendencies are granted by the *Ch'umilal*. The *Ch'umilal* is the sidereal conjunction of the day of birth or fate stars with the day of conception and the future. Future here is understood as what will be the present of a person.

The Barefoot Doctors define four families of *Ch'umilal*. That is, if the total of the *Cholq'ij* is twenty, five of those days characterize each person. According to this definition, each series of *Ch'umilal* define the birth or centre of the person and four auxiliaries who together will give direction to the life of a human being.

> [H]is birth, Nawal is the center, the arms grab the Auxiliaries to have skills that help him in life, his right leg is the one that proceeds forward to go to the meeting of his Destiny, the left is left behind but always follows the person since it is his origin or Conception.[14]

"To be" has a direct impact on the life of the human being in the time count of 260 days. It helps one to know the purpose of one's life, of one's existence. One's ignorance according to this perspective can lead to diseases incomprehensible to positivist science, auditory and visual hallucinations, mood disorders (depression, suicidal impulses, etc.) and sleep disorder.

According to Felipe Pol[15] (a member of the Barefoot Doctors), the treatments that Indigenous specialists follow and recommend are not comprehended from the stand of positivist science. This incomprehension would be due to the differences on what Indigenous specialists and positivist scientists see as the cause of illness. Generally, the K'iche' and Maya perspectives point to socio-economic realities and emotional issues of the family and social environment as the roots of illnesses. Health issues are also related to the inconformity of ancestors with a certain circumstance or disposition, which then manifests as physical, emotional and mental discomfort. And of course, they can be due to physiological reasons as such or combined with emotional and mental discomfort. Hence, treatment requires the unbreakable link between the social, natural, spiritual and material spheres of life.

In an interview in 2017, Felipe Pol narrated the example of a depressive young Indigenous man engaging on self-laceration due to extreme poverty. The man suffered because his family had no land and was thus unable to get a mortgage to emigrate to the United States to "have a better life" thus fulfilling the American dream. Felipe Pol commented that the many diseases people suffer today should be treated by Indigenous medicine, which implies understanding the contradictions of contemporary times. On the one hand, he said, the public is encouraged to participate in food, technology and fashion consumerism. On the other hand, the state and its public policies are designed to perpetuate extreme poverty as a condition for the exploitation of Indigenous peoples. Consumerism and food industrialization produce diseases such as diabetes and hypertension, among others, which did not exist a hundred years ago, so Indigenous medicine must be updated from its own epistemology on its own terms.

Finally, Felipe Pol tells us that the state and its positivist approach to medicine disqualify Indigenous peoples' views on health. However, the state does not guarantee access to the government health system, as hospitals lack medical personnel and basic medicines. For this reason, Indigenous peoples' resistance must be one which deepens their own knowledge and practices, as well as one which strengthens Indigenous people's ability to respond to contemporary needs.

The daily use of other calendars in the first decades of the Spanish invasion diminished the use of the Mayan calendars. In the following centuries this situation continued due to the repression of Indigenous peoples and knowledge by colonial societies and the Catholic Church. It is estimated that six out of eight people who inhabited what would become the American continent died in the first century of the Spanish invasion. Much knowledge and many practices were extinguished, together with the people who possessed them, during the wars and epidemics that came with the Spaniards. Perhaps because of this, or because it always had that goal, *Cholq'ij* also has agricultural and forestry functions. This knowledge is applied to cut down trees to obtain domestic fuel and build houses. The *Cholq'ij* marks the days for the rituals of the seeds that are sown and for the days when the fruits of the earth will be harvested. As indicated by the Barefoot Doctors, specific days are followed for rearing animals and analyzing the appearance and control of pests, among many other uses.[16]

It is necessary to locate the centre of influence of the *Cholq'ij* and its four auxiliaries to determine a harmonic relationship with the plane of the *Wiqanel Tz'uqanel*/form-givers, the builders, the timeless. The timeless are those beings that can move between different planes of existence,[17] for example: Dreams, the life of the people of the past, the future of those who are not yet born, etc. There are several planes. The plane of the *Rajaw q'ij alixak*/the owners of the 13 groups of 20 days that govern the time of human existence. The plane of the *Q'atab'al tzij*/authorities of the different specialties of the health of the spirit, mind and body; authorities of the different manifestations of the beautiful and the tragic; those in charge of delivering justice, among others. The plane of the *Ujer winaq*/the past and the *Qitqa qamam*/grandparents and grandmothers, generations of human beings for which we exist. The plane of the timeless *Majun ke'unatasaj*/the forgotten, the living or dead beings that no one remembers. The plane of *Ri aj paraqan ja*/those who take care of the beams of the buildings, and the plane of those who oversee the *Tzajalinik*/rounds in the day with a frequency of 12 hours, 6:00 a.m. and 6:00 p.m.

> *Chi sacha ri qamak ujer qati't qamam, xix xalek xix paqe'ik chu taq'aj, chu juyub' taq'aj. Xix q'axik, xix ta'ne'ik, xix tojanik, xix tioxinik ruk' mejelem xukulem. Chuch tat, kujitaloq, kujiwilo'q chi toq'ob' le iwuch chaqij. Qati't qamam, ixsutz', ixtew kaqiq' chik. Ixpoqalaj, ixrax amolo' chak. Chi kuyu' qamak, kojkon ta wa' che, xa rumal xa taq oj ik atz'am, oj kinaq' ixim.*
>
> Excuse our faults, Elder grandmothers and grandfathers, you who walked on mountains and on plains. You who lived your life, you who plead for life, you who offered for life, you who thanked for life by placing your knees on the earth. Mothers and fathers, please, look at us, sympathize with us. Grandmothers and grandfathers, you who are now clouds, wind and breeze. You who turned into dust and green flies. Excuse us, we are making a lot of mistakes, but please understand us, we are just made of chili and salt, beans and corn.[18]

Or said another way by the Mayan poet k'ich'e Humberto Akab'al, in his poem "In the Voice":[19]

> In the voices
> of the old trees
> I recognize that of my grandparents
> Guardians of centuries.
> Their voices are at the roots.

Ichek'/dreams

Dreams are an important element among contemporary Indigenous peoples. The colonizing effect through Christian indoctrination from both Catholic and diverse Evangelical denominations (since the second half of the 20th century, mainly the Pentecostal denomination) have led to various logics in the interpretations of dreams. However, dreams are still closely analysed by the interpreters of dreams called *Ajq'ij*/time counters, as part of the practices of the traditional spiritual rituals, which are called *Kotz'i'j*. . The concept of *Kotz'i'j* can be literally translated as moments of flowering and germination of the being. The procedures of the rituals seek the recovery or the continuity of wellbeing through fostering a moment of relief, reflection, dialogue, and harmonization with the social, natural and spiritual environment. For Catholics, the people who interpret dreams are those who are engaged in prayer. For Pentecostals, the people who interpret dreams are the pastor and those who pray. Ironically, both Pentecostal Evangelicals and Catholics secretly turn to the *Ajq'ij*. Until now, there has been little academic research on these contemporary spiritual forms; perhaps for that reason they are free and sovereign to manifest themselves and become energized as everyday social phenomena.

In the first half of the 20th century, the *Ajq'ijab'*/counters of days were the main people consulted regarding dreams. Sleep is considered to be the time of independence from the constitution of living on this plane. The *Jaleb'*/soul or spirit of living beings and the *B'aqil/body* and *Ch'ukul*/body structure can be separated. The body rests on its materiality and the *Jaleb'* can be mobilized in many spaces and transported in multiple temporalities. Awareness of this possibility is determined by the *Q'ij alixak*/day of birth of the person, through how much the person was fed and how much the person does for the collective with their gift.

Dreams are manifested to the *Ajq'ijab'* to prevent them, inform them, show them events of the future, the past, and from another place/time, or someone who will visit them. Complex stories are dreamed of, with characters interacting to form abstract understandings of diseases, and giving messages of all kinds and knowledge from any discipline. In many cases the dreamer participates as a character in the plot or as an observer of it as was the case in the dream I relayed earlier in the chapter. The plots of dreams are metaphors and allusions reminiscent of real or possible facts in multiple perspectives. They are there to understand beyond any one perception and to diagnose malaise, illness, land disputes and demands of the *Ujer qatit qamam*/grandmothers and grandfathers.

Dreams are indicators to know if the rituals necessary for the planting of human life have been fulfilled, because the maternal uterus is considered to be a seedling and to be born is to transplant life. Because the uterus is a seedling the rituals of *Tikab'al q'ij alixik*/sowing life on earth must be performed before the ancestors until the *Solinik k'aslemal*/untying of life leads us back to the seedbed. The rituals of untying life take place when human beings come to have a decadent old age. When babies get sick and there's no medical reason for vomiting, diarrhoea or fainting, someone close to the baby dreams of being given clean water. This is an indicator that the gift of the new life has not been sown to the Earth and that life is thus not yet completely separated from its mother.

In the first weeks of December of 2019, my family had to untie the life of my grandmother María Morales Toj. We asked the time-space to come back into the seed to plant her life inside the Earth. We asked for her to turn into blue flies, flowers, and fruits. This is what happened:

> She was 93. It was a few years already since she lost her memory. She remembered episodes of her life, some cheerful and always those that were very sad. She cried over her dead babies many times and over never having her mother by her side. She remembered the endless beating by her husband, a man she was forced to marry when she was 19. She often thought of the tiredness she felt after days of travel carrying horses with dried chili between mountains and plains when she was a girl. Oh *Wati'tMarya'*/ Grandma Maria! For her living turned into a pain one day. The skin of her hands was starting to come off when someone touched it. Concerned, her daughters would say, it is not possible to suffer like this! They consulted the *Aj q'ij*/counter of days and she recommended the *Solinik k'aslemal*/untie from life to return to the seedbed. Then the family gathered: daughters, sons-in-law, grandchildren and the *Xikin*/great-grandchildren. The mission was to talk about the life of the *Wati't*/grandmother. Yes, said the *Aj q'ij*/ counter of the days, remembering the episodes the family shared with her was necessary to find a pain, a snout, a fact that kept life unleashed… With stunned ears and an appeased soul those present listened to the life story of *Wati't*. She recalled how her stepmother would leave her without food or sent her off for the day's work grazing raw beans. She recalled the times her daughters witnessed powerlessly when the *Mam*/grandpa hit her. He even hit her the day after giving birth. They often heard her curse her father because it was he who decided to grant her in marriage to the *Mam*. – "Ha…! That's it!", said the *Aj q'ij*/counter of the days, we will invoke the name of the *Xikin*/great-grandfather in the thirteen groups of twenties of the *Cholq'ij* to apologize on behalf of the *Wati't*. He must have had his reasons for considering that his daughter should marry him, *k'o rawasil*/as a consequence of continuing with this grudge in the heart. We will never

> know, but in that pain *Wati't*'s life is still tied. And so it was, on the sixth day, the *Wati't* in a quiet dream slept forever.

Despite the diversity of creeds among today's Indigenous people in Guatemala and the civilizing processes that have determined our history, dreams and their meanings remain an important element in everyday life. Luis de Lión in his book *Times Start in Xibalb'a* refers with sarcasm and irony of the racializing that some Indigenous people suffer from in everyday civilizing life thus:

> The Piscoy sings for you and your body shakes. You think something is going to happen to you. But you are indian [racially pejorative term with which Indigenous peoples are called in Guatemala]. Maybe you do not believe in omens anymore. Maybe your head already has other ideas inside, maybe you live in the city, maybe you already know something about the science that is in the books. But if you are indian and you go back to your people and go out at night and you hear the Piscoy singing to you, you forget your city, your books, your science, your new ideas and you say: – I believe in God and not in you – but you believe, you sign yourself with the cross and for many days you are awaiting to see what is going to happen to you. Maybe nothing will happen to you, maybe what happens to everyone will happen to you on your foot or hand, but by sheer chance and without major importance, a fight in the canteen or the same fear that something is going to happen to you, anything that happens, you will say it was because of the Piscoy.[20]

Rawasil/the warning message, the reaction of the action

Dreams and dreaming are very related to the *Rawasil*/warning message or reaction to some action or thought. A discomfort caused by the break of the harmony with something or someone who might be alive on this plane or absent from it and rather present in another space-time. People get sick, or worse, the people they love the most get sick or suffer. Others suffer from depression, have episodes of apparent madness, have headaches or body pains without reason or cure, money does not come easily, it gets stolen. The *Ajq'ijab'*/counters of days and the *Iyomab'*/midwives, when they care for a patient investigate and think about what did the patient say? What did the patient do? The illnes *Xto rawasil*/is a consequence of the patient's action or thinking.

Ajq'ijab'/counters of days and the *Iyomab'*/midwives, diagnose people who needs treatment and identify the plane of time/space where the treatment is needed. This is why the patient's dreams are important and why the *Ajq'ij*/counters of days keep all their support tools open. When you dream, your joints speak to you. The blood has special ways of manifesting itself anywhere in the body to say what the problem is and why it emerged… Hence the *Tz'ite'*/sacred wrap composed of *pito* tree seeds is consulted to say what is wrong with one's body and to look for answers to the problem in the different planes of existence and in different times. The wrap searches for the right treatment looking at what was there before and to whom it belongs.

The Tz'ite'/sacred wrap is composed of 260 seeds of the Pito tree. The *Pajanik*/statistical approximation has games of ones, twos, threes and fours. Their possible combinations would tell truths, lies, sadness, desires, debts, illnesses, faults and dreams... These narratives that show the combinations of the seeds in subsets of ones, twos, threes, fours are formed in the disintegration of subsets of four – of sets of seeds that are separated randomly from a total of 260 seeds. To this set of 260 seeds, a person explains the problem or question they have in order to ask for help from the set of seeds. Each seed represents each of the days of the *Cholq'ij*/Lunar calendar. They, the *Tz'ite'*, can see, hear, and know what happens, what is being thought, what is being said, what was said, what one suffers, what is needed by the various ways of being in space/time. For this reason, the set of seeds of *Tz'ite'* chosen for this function are called a rod. They constitute a local measurement system because with them a measurement or an approximation is made, and a solution is given to the patient.

Many Indigenous people in Guatemala consider themselves active members of the Catholic or Evangelist Churches. Yet others consider themselves to be skeptical of older beliefs like the *tz'ite*, but when the *tz'ite* or dreams say something about these skeptical Indigenous people... then they do worry and say "something must be done," and *Chimaje'* takes place...

Chimaje'/It is like that because that is like it is/like it should be done

Chimaje'/It is like that because that is like it is, is the exercise of certain practices that are part of the daily life of the contemporary Maya still living a communitarian quotidianity. In large cities the idea of collectivity is diluted in the time spent in traffic, in meetings scheduled, in survival, and in superfluous consumerism.

> An evangelical pastor of Argentine origin from the Pentecostal slope, Jorge Rojas, bought a plot of land and built his house in *Chulumal*, a village in Chichicastenango, Guatemala. He said that, for his effective insertion as a neighbor of the community, he had to serve as a member and secretary of the Water Committee. This community service is to ensure local water tubing projects. In his time serving as secretary, the Committee had to integrate a new source of water into the existing tubing project. The chairman of the Committee was also an evangelical but an Indigenous Maya K'iche' pastor. At a project expansion planning meeting, the Committee's president included in the agenda the appointment of an *Ajq'ij*/counter of days to perform the necessary rituals to ask Mother Earth's permission to pipe and use the water from the source. Jorge Rojas said he was surprised and curious so he asked: Why do an ancestral ritual to intubate a water source and not do an evangelical prayer or ceremony, since there is an

> evangelical majority between the users and the members of the Water Committee? The chairman of the Committee only replied: *Chimaje'*/it is like that because that is like it is, that is how it should be done.[21]

This situation illustrates the contemporary perception of the practical function of ancestral rituals. Their "whys" and "for what purpose" are still confusing in current times among Indigenous peoples who, despite religious membership, rely on these rituals as a way of being respectful to the different planes of time-space which one learns to believe in.

The religious membership of Indigenous peoples in these latitudes are complex when it comes to understanding dreams, remaining in good standing in relationship with the land and its resources for life, with the ancestors, and to help solve economic problems or incurable diseases incomprehensible for positivist doctors. *Kotz'i'jj*/moments of flowering and germination of the being (see page 11) is something that is believed and practiced freely by those of us who publicly subscribe to this way of doing/believing. For others, Catholics and Evangelicals who secretly visit the *Ajq'ij*/counter of days, *Kotz'i'jj* is lived as a discreet practice. Regardless of believing in a Catholic, Evangelical, or Mormon religious discourse, people's sociocultural reality is rooted in the Earth, the ancestors, and dreams.

Due to these complex validities of historic-cultural feelings, knowledges, and makings, Christian religions have complex and diverse amalgamations in Latin America.[22] Hence people who live in small towns and live together in contemporary Indigenous communities, reproduce practices and conduct themselves by precepts in the memory of *Chimaje.'* It is known that well-being, counting with enough natural elements, and maintaining life must be a matter of seeking harmony and the manifestation of gratitude. The earth, the water, the fire, the air, every element that helps us maintain life, is life itself and not a god or a goddess as stated by "ethnoscience" from a stand of alterity. Unfortunately, the rituals in contemporaneous times are momentary and sporadic, and are fewer each day in the practices of community life. Community organizing and norms are, in the present, a key possibility for community survival in Mayan quotidianity.

The presence of the concept of *chimaje'* in the everyday lives of Indigenous peoples is currently threatened by the expansion of the idea that reality is mass production, overuse and economic survival. Economic survival suffocates people's lives. *Chimaje'* supports wellbeing in the community living side-by-side. A concept to define an ancestral principle of conviviality and coexistence is the *Chalaj juk'ula*/existing side by side.

Chlaj juk'ula/existing side by side

Yojatajik[23]

> Iwir,
> ri jab' xusipaj kan utzil wochil pataq qanima'

xujopij kan kotz'i'j chuwoch kajulew
xusutij kan chachal re k'aslemal pataq le juyub' taq'aj
jexutzij kan roxox re ja'maril pa taq ri komon xo'ral.
Iwir,
ri uwa'ral jab' xutzij ub'oq'och
xuxlan chi' le b'e
xuquje' xtze'tzatik are xrilo xe'tz'an ri ak'alab'
ri eq'aq'al chikop
je ri ech'umil.
Kamik,
le jab' kut'oqopij b'ik kiqul le juyub' taq'aj
je kuch'orerej loq kaminaqib' choch taq qaq'otz.
Kamik,
le uwa'l jab' kujumu'ba'
kujuq'utuq'a'
je kujulik'ib'a choch le q'ij
rech kaqapakuj b'ik we moya qawaram
je we qayojtajik re Xib'alb'a.
¿Jasche xujyotaj
kuk ri uk'u'x kaj ri uk'u'x ulew?

Disharmony

Yesterday,
rain donated well-being to our hearts
scattered flowers on the face of the cosmos
circumnavigated necklaces of life in the mountains and plains
and lit peace roses in the communities.
Yesterday,
rainwater lit the eye
rested on the side of the road
and smiled when seeing the kids play
to be fireflies
and stars.
Today,
rain tears the neck of the mountains and plains
and drains the dead over our cheeks.
Today,
rainwater soaks us
it rubs us
and extends us to the sun
to clear our blind dream
and our hellish disharmony.
Why do we disharmonize
with the heart of heaven and the heart of the earth?

Anima' re ja'maril[24]

Waral pa rixq'eqal Xib'alb'a,
are taq xchuplin ri usaqil jun q'aq'al chikop,
sqil apan choch le kajulew jun ajwaja'
nab'e kujosq'ij na
kutolob'a' na
je kujam na ri ujuyub'
ataq'aj
je ri utikob'al,
k'ate k'u ri' kapoq'loq
kakotz'i'jan loq
je kawochin loq ri utiko'n.
¿Jas kqbab'na na rech kujux kitikob'al
ri Ajawixelab' re ri junelik k'asilik k'aslemal?
Ajtijab' rech ri q'ijirik k'aslemal
chujtijob'ej na aläq
che ri kesaxik b'ik ri uk'ixal makaj pataq we quxlab'
che ri resaxik b'ik ri ch'ob'onel ajaw pataq we qab'aqil
xuquje' chech ri ub'anik nimaq taq Q'ij chikech we qanima'.
Rech utz, waral – kamik,
ruk' suk' ch'olch'oj qach'umilal,
kqachakuj n ri já'maril re ri k'aslik k'aslemal
kqatz'uk na ri pach'un ib' re ri k'aslik k'aslemal
kqawexaj na ri utzil wochil re ri k'aslik k'aslemal
kqaloq'ob'ejna ri uwoch kajulew re ri k'aslik k'aslemal.

Heart of Peace

Here on hellish dusk,
when it lit the light of a firefly,
we saw towards the face of the cosmos a farmer
first clean
empty
and vacates their mountain
their plain
and their seed bed,
then it germinates
blooms
and fruits their sow.
What do we have to do to become the seedbed
of the Creators of the everlasting life awake?
Teachers of solar life
educate us still
to remove the thorn of sin from our breaths

to remove the rational animal from our bodies
and turn our hearts into great suns.
For good, here and now,
with our perfect and immaculate star,
we work the peace of life awake
we braid the reconciliation of living life
we sow the well-being of awaken life
we value the universe of living life.

Chlaj juk'la, this concept does not have a literal translation, as it explains a very broad principle of coexistence. The *Ajq'ijab'*/counters of the days in Chichicastenango – Guatemala, when they coincide in some places like on a sacred altar, they greet each other by saying and answering *Chlaj juk'la*. This concept is composed of two terms: *Chlaj*, which means accompany, and *Juk'la*, which means two/pair. The concepts speak of the need to be aware of difference, which does not mean the inferiority or superiority in importance nor the contribution or responsibility of someone or something in the universe. The concept applies to the relationship between humans, plants, minerals and animals. It applies to everything that lives a short lifetime like an ant, or a long lifetime like minerals. The example that the elders mention to explain this concept is the left and the right feet, for example. One would not move forward if one had two right feet, even if one had two left feet. The only way forward is if there is a right foot and a left foot. This principle encompasses the pursuit of the harmonious and equitable relationship with beings in the different planes of time/space known to Indigenous people.

There is no perfect harmony nor perfect disharmony, as the poems quoted at the beginning of this section show. In the time and space conception of Mesoamerican Indigenous peoples, nothing exists that is completely good or bad. Living is to learn that goodness has doses of imperfection which are the only possibility of knowing it. Getting it wrong is the only way to learn wellbeing, because if the discomfort is not known, then wellbeing could not be known.

The learnings of *Chlaj juk'la* are not possible for the human condition as an experience. However, they are a permanently desirable condition. When Pablo Garcia says:

> Yesterday, rainwater lit the eye, rested on the side of the road, and smiled when he saw the children, playing to be fireflies and stars. Today, rain tears the neck of the mountains and plains and drains the dead over our cheeks.

This tells us of a contemporary reality in which the neoliberal capitalist system tears apart the life of life… Opencast mining, oil wells, African palm and avocado deserts, pro-transgenic seed laws, industrialized food deposits, bottled water, are examples of the commodification of life. This is the rain that drains dead and living dead that Pablo Garcia describes.

The possible world and life in our human future is bleak if we do not make *Chlaj juk'la* a principle of coexisting. Human life and the permanence of human species is a human responsibility. Life on Earth is dynamic, soon to end – it resurfaces, it adapts, it becomes again. How long humanity can remain on Earth as a species is a question with frightening answers. There is no need for foreboding powers to know that. The omens are many. What we do in this present is important, for on it depends the future.

This writing did not follow the canons of the academy approved as objective, because then these reflections would be constructed from a vision of otherness for otherness. Although the original chapter was written in Spanish, the narrative follows a dialogue around firewood on fire. Those moments are now less frequent because of distant relationships through social networks, smartphones, television or the tiredness of the price of the success of survival in times of fierce consumerism. To this point, as Huberto Akabal said:

Ojer bix re ri kik'el[25]
Man xintu' ta ri' kaxlan tzij
are taq xinalixik
Ri much'abal xalan chi uxo'l k'iche'aj
rumal ri' k'o una'il ulew chi uxo'l;
ri kich'abalil ri wati't numam are ri', ri wachoch.
Are we kinich'awik pa kaxlan tzij,
xa je ta ne kinkoj jun k'ak'a' lawe
ki kutur chi jun uchi'ja re jun k'ak'a ulew
ri jawi' ri tzij k'o wi chi ri kakibij
k'o wi chi ri kakina'o che ri ulew.
Ri kaxlan tzij kuna'tisaj jun k'ex,
are wa man k'ix ta wib kintzijon chupam
xa rumal loq'om wa'
ruk' ri kik'el re ri nuxe'tayil ri uwi' ri nujolom.
Pa ri jun k'alk'a ch'abalil
kink'utu ri uk'otz'ijal re ri nubix.
ri uki'al re k'o wi chi taq bis
ri uwachibal re k'o wi chi taq ki'kotemal…
Ri kaxlan tzij jun lawe chik
che ubixoxik ri ojer bix re ri nukik'el.

The Blood's Old Song

I didn't suck the Spanish language in
when I came into the world.
My tongue was born among trees
and tastes like earth;
My grandparents' tongue is my home.
And if I use this language that is not mine,
I do it as the one who uses a new key
and opens another door and enters another world

where words have another voice
and another way to feel the earth.
this tongue is the memory of a pain
and I speak it without fear or shame
because it was bought
with the blood of my ancestors.
In this new language
I show you the flowers of my song,
I bring you the taste of other sadness
and the color of other joys...
this language is just one more key
to sing the old singing of my blood.

A note on translation

This text was dreamed in Kaqchikel, written in Spanish, and translated into English. Special thanks to Gioconda Coello and Ligia López for all your work and patience to make this chapter a reality in English. For all the learnings and the heir incredible sensibility, many thanks and admiration.

Notes

1 Nixtamalizar is the cooking of dried corn using cal or ash. Cal is calcium oxide (CaO) drawn from limestone also known as "living rock." The *nixtamalizar* process produces complex chemical reactions that make dried corn nutritious for humans. This is a legacy of Indigenous women's discoveries and experiments for centuries, which are now a legacy of women's science.

2 Tortillas are flat cakes made of *nixtamalizado* corn. They are the main source of carbohydrates and amino acids in Mesoamerican Indigenous people's diet.

3 Feminism as theory has been an ideologic pillar to think and rethink gender roles in the patriarchal system. Feminism's contributions have unquestionably contributed to the struggle for women's rights. My critique to feminism here is particularly towards the wave that addresses the "liberation" of women disregarding cultural and historical diversity beyond academic and work spaces. Self-critical reflection must be fundamental today. In times of labour exploitation (whether paid or poorly paid labor) and incentives for voracious consumerism, "agricultural" and "domestic" spaces must be reappropriated as spaces of independence from the global market. Cultivating food for the reproduction of collective life, making food, creating or recreating recipes, for example, are rebel and revolutionary acts against the commodification of food. Agricultural and biodiverse spaces, and domestic spaces of personal and familial self-preservation, are not patriarchal. On the contrary, they are a space that patriarchy has controlled for centuries. Both women and men must redefine these spaces. After all, the reproduction of biodiversity and life happen there, in the fields, the kitchen, and the home. Without reflection, feminism can become aimlessly a tool for commodifying the lives of women if women's liberation from oppression only means their labor in academic and public life.

4 In this chapter, "Millenary Mayas" refers to the originary Mesoamerican populations who, before 1421, had similar cultural, linguistic, agricultural, and nutritional practices and who are the predecessors of today's Indigenous peoples in the region.

"Mesoamerican Indigenous peoples" are understood here as originary populations framed within colonial dynamics, and redefined geographically and politically by the conformations of the nation-state of what is now Mexico, Guatemala, Belize, Honduras, El Salvador and Nicaragua. These are Indigenous peoples who—despite the plagues, forced labour, territorial mobility and ideological control policies through Catholic Christianization—resisted and reconfigured themselves socio-culturally for their own survival until the start of the regional Internal Armed Conflicts and the so-called "green revolution" in Guatemala. By "contemporary Mayas" I mean the Indigenous peoples in Guatemala who lived through the Internal Armed Conflict (1960–1996). It refers to the generations that survived the conflict. The reconfiguration among Indigenous populations—one that encompasses sociocultural, economic, ideological, and political aspects—was determined by the Internal Armed Conflict and the Green Revolution as well as by the proliferation of Evangelical churches, the radicalization of a sector of the Catholic Church, and the forced and voluntary mobilizations within the territories already defined as "national" and "international." The political, economic, cultural, and technological events from the second half of the 20th century in heterogenous ways revitalized, and in some cases, racialized the discourses from and about the ethnic. These events similarly characterized religious and spiritual practices, economic policies, belonging and occupation of territories among other aspects of life.

5 The Pop Wuj has been widely studied as a cosmogonic and historic document. However, there are not yet endogenous analysis by Indigenous women, of the establishment of current patriarchal structures in the stages that the document describes. The oral history (which is being compromised in current times) by which many of us had the privilege to learn the histories of the Pop Wuj, tells of an era previous to the "first men." In that previous era, women owned scientific knowledges. Women had power over/with plants and animals. They were heads of households, had deciding power over political matters, for instance. The most well-known among there were Ixmukane and Ixkik.

6 Adrián Inés Chávez translated the Pop Wuj for decades, as he himself indicated in the book's introduction. In those many years, he created a system of K'iche' symbols which he proposed as an official system for the language. The book is then written in four columns, the first column is a literal translation of the manuscript credited to Fray Franciso Ximenez. The second column is a transcription of the manuscript with the symbols Chávez created for the K'iche' language. The third column is a literal word by word translation into Spanish, and the fourth column includes a grammatical Spanish translation. The texts cited in this chapter come from the fourth column. It was not possible for me to transcribe the text from the second column given that the symbols Chavez created for the K'iche' language do not exist digitally.

7 Adrián Inés Chávez. Pop Wuj. Poema Mito Histórico Kí-ché. Quetzaltenango: Centro Editorial "Vile", 66–67, 1978.

8 Adrián Inés Chávez. Pop Wuj. Poema Mito Histórico Kí-ché. Quetzaltenango: Centro Editorial "Vile", 66–67, 1978.

9 Médicos Descalzos are an association of Indigenous women and men, midwives, bone specialists, traditional doctors, family counsellors, spiritual guides, among other experts, that meet every first Friday of the month at the City Hall in Chinique de las Flores of Quiché state-Guatemala. They have been doing so for 22 years. The objective of their meetings is to focus on sharing knowledge and treatments to meet the needs of their patients.

10 Asociación Médicos Descalzos. *Yab'il xane K'oqil? Illnesses or Consequences? Six psychopathologies identified and treated by Maya'ib K'ich'eib' therapists*. Guatemala: Editorial Cholsamaj, 38, 2012.

11 Each day has a specific meaning in different aspects of human, animal, mineral life, says regarding the elements, the movements of the body, the planet, the sky, etc.

12 Nawal is a category that summarizes the historical concept of being in space/time, according to the contemporary Mayan. The term nawal has no meaning in many of the Mayan languages, and if it does, it does not have a direct relationship with the lunar calendar system. Nawal is a concept that was introduced in the Mayan languages by specialists and scholars, mostly non-Indigenous, studying the Maya. As such Nawal has been mediatized by so-called Mayan intellectuals and NGOs leaders.
13 Asociación Médicos Descalzos. *Yab'il xane K'oqil? Illnesses or Consequences? Six psychopathologies identified and treated by Maya'ib K'ich'eib' therapists*. Guatemala: Editorial Cholsamaj, 39, 2012.
14 Asociación Médicos Descalzos. *Yab'il xane K'oqil? Diseases or Consequences?*, 41.
15 Felipe Pol, personal communication, October 2018.
16 Asociación Médicos Descalzos. *Yab'il xane K'oqil? Diseases or Consequences?*, 55.
17 In K'iche' nouns do not have gender like they would have in Spanish. They have specific nouns to designate feminine and masculine elements that must be named to define gender. Thus, form-givers and builders have no gender.
18 Sebastián Chumil (also known as Juan Xón), K'iche' prayer, 7 of February 2003.
19 Humberto Ak'abal. Roqonch'iaj. *Cry*. Kaqulja K'ich'eib Series. Guatemala: Editorial Cholsamaj, 62, 2004.
20 Luis de Lión. *The main time in Xibalbá* (Antigua Guatemala: Ediciones el Pensativo; 2012), 109.
21 Personal Communication, Jorge Rojas, October 2018.
22 Jean-Pierre Bastian. Para una sociología del cambio social en la modernidad periférica [Towards a sociology of social change in the peripheral modernity] (México: FCE, 1997).
23 Pablo García. B'ixonik tzij kech juk'ulaj kaminaqib'. Canto palabra de una pareja de muertos [song for a dead couple]. Guatemala; F&E, 40–41, 2009.
24 Pablo García. B'ixonik tzij kech juk'ulaj kaminaqib'. Canto palabra, 70–71.
25 Humberto Ak'abal. Ri Tzij Kek'iyik. Las palabras crecen [Words grow]. Guatemala: Editorial Maya Wuj, 16–19, 2010.

References

Ak'abal, Humberto. *Roqonch'iaj. Cry. Series Kaqulja K'ich'eib'*. Guatemala: Editorial Cholsamaj, 2004.

Ak'abal, Humberto. *Ri tzij kek'iyik. Words grow.* Guatemmaya: Editorial Maya Wuj, 2010.

Asociación Médicos Descalzos. ¿Yab'il xane K'oqil? ¿Enfermedades o Consecuencias? Seis psicopatologías identificadas y tratadas por los terapeutas Maya'ib K'ich'eib [Yab'il xane K'oqil? Illnesses or Consequences? Six psychopathologies identified and treated by Maya'ib K'ich'eib' therapists]. Guatemala: Editorial Cholsamaj, 2012.

Bastian, Jean-Pierre. *Para una sociología del cambio social en la modernidad periférica [Towards a sociology of change in the peripheral modernity]*. México: FCE, 1997.

Chavez, Adrián Inés. *Pop Wuj. Poem Historical Myth Kí-ché.* Quetzaltenango: Editorial Center "Vile", 1978.

Garcia, Pablo. Canto palabra de una pareja de muertos. *B'ixonik tzij kech juk'ulaj kaminaqib'. I sing word from a couple of dead.* Guatemala: F&E, 2009.

Pol, Felipe. Personal communication, October2018.

4 Spirits and serpents

Buddhist prosperity in the 'Snake Temples' (Mway Paya) of Myanmar

Nicole Tu-Maung

Snake Temples (*Mway Paya*) are novel Theravada-Buddhist places of worship in Myanmar that are made notable through the veneration of live, captive Burmese pythons (Python *bivittatus*). The pythons who inhabit these temples are viewed as *nats*, spirits of a pre-Buddhist animist tradition that is considered Indigenous to present-day Myanmar. The *nat* spirits are specifically those associated with the *Naga*, a powerful serpent of the spirit realm. Embodied by living snakes, these spirits are perceived by Buddhist followers as sacred entities embedded within, or indigenous, to particular landscapes upon which Snake Temples are constructed.

At Snake Temples, followers pay respects to the serpents as a means of affirming their bonds to the Buddhist religion. Through ritual and worship, the faithful accrue karmic points which are expected to provide positive returns in future lifetimes in the cycle of birth and rebirth. Strengthening bonds to the spirit world also provides devotees with the opportunity to seek supernatural intervention in the mortal realm to pursue wealth, health, and material goods to achieve a future that they desire. At the same time, veneration of guardian spirits helps fulfill the central duty of Buddhist followers to continually protect and propagate the Buddhist religion so that it may flourish into the future. Snake Temples are found in the peri-urban areas of Yangon and Mandalay, Myanmar's two largest urban centres. The followers who worship at Snake Temples are diverse members of the Buddhist laity, including members of the rural poor, the urban elite, and the politically powerful, all striving to achieve an idealized vision of the future: For themselves and for the religion. Through their incorporation into contemporary Buddhism, new future(s) are produced for *nat* spirits, who become entangled in novel Buddhist discourses and places of worship.

Though heavily influenced by long standing animist traditions and said to house ancient *nat* spirits, Snake Temples were first built into Myanmar's religious landscape during the early 1980s,[1] a time of rapid economic and political change. Beginning in this period and continuing through the millennium, Myanmar experienced a transition from financial collapse to the 'opening up' of capitalist markets alongside changes in government leadership,[2] growing secularism,[3] and heightened conflicts with non-Buddhist groups,[4] all which create turmoil in the Buddhist cosmological framework.[5] According to religious historian, Niklas Foxeus, from the 1980s to the present day, there was the emergence of "prosperity Buddhism" cults in Myanmar, which strongly emphasize veneration of *nat* spirits as a means to achieve personal and economic security.[6] Simultaneously, veneration for *nats* within a Buddhist context allows followers to fulfill their duty to propagate and perpetuate the religion, which is perceived to be in decay. He argues that novel forms of Buddhist worship, centred on the acquisition of economic and religious prosperity through spiritual intervention, help the Buddhist laity to moralize material desires, make sense of unforeseen transitions, and gain ontological security in future(s) of great secular and existential uncertainty.

In this chapter, I present the first ethnographic account of Snake Temples, which are based on observations and interviews that I conducted between 2018 and 2019. Based on my research, I argue that through these temples indigenous *nat* spirits and their believers produce future(s) for themselves within novel Buddhist imaginaries and sacred spaces. These future(s) are linked to a particular form of 'prosperity Buddhism,' one that is created by a community of humans, spirits, and snakes. Importantly, these temples create spaces in which indigenous *nat* spirits persist amidst the social dislocations of capitalism, and in doing so offer believers a more tangible, stable, and self-determining hold over their future(s).

In the first section, I explain some of the complexities associated with the concept of Indigenous as it is used in Myanmar as well as a detailed explanation of my use of the term indigenous to refer to the spirits, beliefs, and cosmologies which take place in Snake Temples. In the second section, I draw on the work of Niklas Foxeus on 'prosperity Buddhism' to contextualize Snake Temples within a larger Buddhist movement. By building on Foxeus' research, I show that *nat* spirits are important to spiritual learning(s) and the creation of desirable future(s) in the context of capitalism and globalization. In the third and longest section, I present ethnographic details and oral histories from three different Snake Temples. Through each of the sites presented in this chapter, I analyze the ways in which the incorporation of indigenous spirits into Buddhism produces futures(s) for humans and spirits on three scales: 1) futures for the spirits through the future of the Buddhist religion and its sacred landscapes; 2) imagined futures of economic and material security of Buddhist devotees in a capitalist society; and 3) possible futures for political actors within a sociopolitical system based in Buddhist ideologies.

A note on Indigeneity in Myanmar

International legal definitions of Indigeneity, such as the United Nations Declaration on the Rights of Indigenous Peoples (UNDRIP),[7] serve to support colonized, underrepresented, and oppressed minority groups in the areas of human rights,[8] cultural rights,[9] and rights to land or territory on the basis of historical attachments to place.[10] Though Myanmar is signatory to the

UNDRIP, international concepts of Indigeneity are strikingly incongruent with localized meanings of Indigeneity within the context of Myanmar.[11] In Myanmar the term 'Indigenous Peoples' and ideas of Indigeneity have been incorporated into the national constitution under the term *taingyintha*,[12] a category used to refer to ethnic groups which are recognized by the government as belonging to the 'national races.' The analytical utility and ethics of the term Indigenous is contested in the case of Myanmar because the concept of Indigeneity as it applies to peoples has been used by the state to exclude certain ethnicities, such as the Rohingya, and justify political violence against those who are not in Myanmar's official list of 'national races.'[13] It is not within the agenda of this chapter to use the term Indigeneity in reference to groups of people or ethnicities in Myanmar, but rather to explore how notions of indigeneity apply to spiritual beings and beliefs. Throughout the chapter I capitalize the term Indigenous/Indigeneity in reference to people who identify themselves as such. I use the term indigenous with lower case to refer to spirits and their embodiment through snakes. I make this distinction with consideration of the contested politics of indigeneity in Myanmar, and as to not suggest that the experiences of *nat* worship communities are necessarily comparable to that of Indigenous peoples globally.

Scholars of modern Myanmar often refer to animist *nat* worship as a set of practices and beliefs, or religion, that is indigenous to present day Myanmar, as in existing prior to the introduction of Theravada Buddhism in the 11th century.[14] In Western scholarship, discussions about Indigenous religions are often juxtaposed against 'world religions' or 'new religious movements' as an analytical category that is distinct from mainstream world religions.[15] However, my discussion of Snake Temples will challenge indigenous/mainstream religious dichotomies by illustrating ways in which elements of *nat* worship have been absorbed, integrated, and transformed by Buddhism, producing future(s) for indigenous spirits within Buddhist cosmologies and spaces. I will therefore be examining elements of indigenous *nat* worship through the lens of contemporary Theravada Buddhism. Through this examination, I am not suggesting that Buddhism is an Indigenous religion to present day Myanmar, nor do I seek to contribute to Buddhist-Nationalist arguments in favour of Buddhist majoritarian or "nativist" politics, a narrative which has been used in Myanmar to legitimize the social and legal oppression of non-Buddhist minorities within the country.[16]

Scholarships of Indigenous religions as a category are also frequently coupled with discourses about Indigenous peoples.[17] Such a coupled approach contributes to the reproduction of stereotypes about limited beliefs held by specific peoples and creates assumptions about uniformity across spans of time, space, and lineages.[18] The analysis of Indigenous religions as coupled with Indigenous peoples can be limiting to critical scholarship, which is especially true in the case of Myanmar.

My use of the term indigenous in reference to Snake Temples is two-fold. I use indigenous to describe the ways in which followers of Snake Temples perceive *nat* spirits as being endemic or embedded within particular landscapes and with whom they have an ancestral bond.[19] Though these temples were constructed during the late 20th century, those who worship at these sites perceive the *nat* spirits to have existed as part of the landscape for millennia and expect that they will continue to reside there for eternity. Simultaneously, I refer to *nat* worship as a set of practices which predates Buddhism, or is indigenous to the geographic region of Snake Temples, but that are decoupled from any particular peoples or ethnicity. I instead analyze ways in which *nat* worship is incorporated into contemporary Buddhist cosmologies. Those who worship at Snake Temples identify as Buddhist, however they represent various ethnic groups including Burmese, Mon, Shan, and others. Followers at Snake Temples are primarily those from central or southern Myanmar, urban and suburban residents of areas nearby these temples. Religious tourists from other parts of the country also visit these sites, thereby expanding the geographic reach of these indigenous concepts.

Snakes Temples and the 'Cult of the Guardians of the Treasure Trove'

In the Snake Temples of Myanmar, *nats* and Buddhist followers work together and communicate to produce future(s) for both humans and spirits, existing within a Buddhist cosmology. I situate these sites within the context of a larger 'prosperity Buddhist' movement by building on the work of Niklas Foxeus, a religious historian of contemporary Buddhism in Myanmar.[20] His research unfolds the narratives and practices which surround a similar suite of novel temples surrounding Mandalay and Yangon which he calls "the Cult of the Guardians of the Treasure Trove," temples which he argues represent a form of "prosperity Buddhism."[21]

According to Foxeus, 'The Cult of the Guardians of the Treasure Trove' consists of non-institutionalized, novel Buddhist temples which are predicated on the existence of magical, sacred treasures which enliven particular landscapes within the Buddhist imaginary.[22] His research concludes that narratives surrounding these temples converge on the shared belief by followers in the existence of *thaik*, or treasure trove. *Thaik* are sacred objects perceived by communities of Buddhist followers to exist within the physical landscape at particular sites. These objects typically include ruins of ancient Buddhist temples or relics of the Buddha and possess magical Buddhist qualities or *dago*. Temples within 'the Cult of the Guardians of the Treasure Trove' emphasize the worship of local *nat* spirits who are viewed as guardians of these magical treasures and landscapes. In Foxeus's research, he finds that Buddhist followers who frequent these temples choose to do so because of a perceived personal connection, known as a *thaik-hset*, to the treasure trove and its guardian spirits. He highlights that followers choose to worship at these sites in order to strengthen their bonds to the *nat* spirits so that they can seek supernatural help in gaining personal prosperity in the form of economic or personal success.

He concludes that novel temples that share similar narratives constitute a form of 'prosperity Buddhism' which developed and gained popularity during the growth of the capitalist economy of the late 20th century.

Based on my own research, I argue that Snake Temples represent a unique subset of Buddhist sacred sites which should also be considered a form of 'prosperity Buddhism.' My research also demonstrates the syncretism between *nat* worship and contemporary Buddhism in Myanmar, where *nats* are conceived as existing within Buddhist frameworks while also interacting with the sociopolitical and economic landscape. I highlight important, varied functions that *nat* spirits serve, such as protecting sacred spaces within the suprahuman Buddhist realm as well as aiding humans in their pursuit for their desired future(s). My research also reveals the application of spiritual learning(s), such as communications with spirits through dreams and possessions, to informing activities such as the construction of sacred sites, the formation of local economies, the negotiation of political power, and the creation of communal discourses. In this chapter, I decouple the analysis of indigenous spirits from specific groups of people or periods of time. At Snake Temples, *nat* spirits are indigenous in that they are tied to geographic spaces in the landscape. However, the worship of these spirits is not limited to any particular group of people nor time. Instead, indigenous *nats* are points around which Buddhists of various ethnicities and socioeconomic positions intersect through common beliefs, practices, and discourses.

Through interviews[23] with dozens of Buddhist monks, lay people, and temple committees (*gawpaka*) at six different Snake Temples near Yangon and Mandalay, I have found that narratives of *thaik, thaik-hset*, and guardian *nat* spirits described by Foxeus are also central to religious discourses surrounding Snake Temples.[24] However, I assert that these sites are distinguished by the incorporation of *nats* as live Burmese pythons, adding a unique multispecies dimension to the 'prosperity Buddhist' movement. Importantly, the prosperous future(s) of beings from spiritual and physical worlds are negotiated through ritual and religious discourses. Devotion to *nats* in the form of live Burmese pythons opens avenues for followers to overcome the uncertainty of the future (s) they desire. The presence of the pythons serves as a form of physical evidence that *nat* spirits exist in the landscape. Snakes are the living, breathing, biological entities in which the otherwise elusive spirits take place.

Similar to practices at temples within the 'Cult of the Guardians of the Treasure Trove,' communication with spirits at Snake Temples takes place through donation rituals (*nat pwe)*, the interpretation of dreams, and spirit possessions (*winn-puu*). In other words, *nat* spirits provide the Buddhist laity with supernatural learning(s) such as knowledge about sacred landscapes and actions that individuals and communities should take to create positive future outcomes.

In this chapter, I discuss communication with *nat* spirits as learning(s) that are facilitated by dreams and possessions. My interpretation builds on the work of Thomas Patton, who has written on the importance of 'phantasmagorical' dreams in the Buddhist imagination.[25] He writes, 'for Buddhists in Myanmar, rather than being sources of suspicion and disregard, dreams are fonts of visionary knowledge to be taken seriously and respected. They are foundations of omens and divine revelations and are part of a suprasensible world that has an ontological status, one that is as real as the one they perceive with their senses.'[26] I add to Patton's argument by documenting the histories and narratives surrounding Snake Temples, where dreams are important sources of information through which a community of believers learn about the sacred world in order to guide individual and community decisions.

Creating future(s) in the Snake Temples of Myanmar

Through the incorporation of *nat* spirits into Buddhist sacred sites, rituals, and cosmologies at Snake Temples, they come to bear important meanings and functions for communities of followers. Future(s) and ongoingness[27] are produced for the indigenous *nat* spirits who exist in both the sacred geographies of the Buddhist imaginary and in the temples of the built environment.

Importantly, the *nats* which are venerated at Snake Temples are local *nats*, bound to specific geographic sites. This contrasts from the veneration of the Cult of the 37 Nats, an official pantheon of *Nats* that is more broadly recognized by Buddhists in Myanmar.[28] At Snake Temples, local *nats* are considered indigenous to particular geographic sites and have unique, novel identities. Some *nats* at Snake Temples have unique names including *Narawata Mehdaw,*

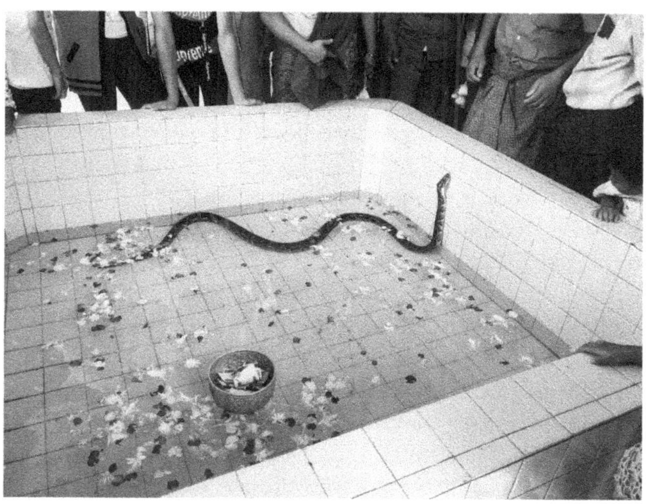

Figure 4.1 At a Snake Temple near Mandalay, followers observe a Burmese python during the daily bathing ritual. As the snake swims, a temple official (*paya lugyi*) recites Buddhist incantations (*paritta*), which are expected to bring positive returns for those who place donations into the floating, silver alms bowl.

Yeh Lei Thuzar, Saw Nan Nway, and others. Some *nats* are referred to only by generic or descriptive titles such as *A Ko Daw Gyi*, meaning Respected Elder Brother, or *Chaw Chaw A Ma Daw*, meaning Beautiful Respected Sister. Individual snakes residing at the temple are referred to by the names of the local spirits they embody, a tradition which instills agency, identity, and power upon the non-human actors of these multispecies spaces.

By presenting ethnographic vignettes and details from local discourses of three Snake Temples, I will present the ways in which Buddhist followers communicate with and learn from indigenous *nat* spirits to create novel Buddhist spaces and attempt to achieve various future(s).

Making sacred space on the Mandalay-Lashio highway

The winding road towards Pyin Oo Lwin offers ghostly views of half-barren hillsides, scoured in highway expansion projects and limestone extraction. Amidst this rapidly changing landscape, a Snake Temple on a modest two acres stands on a small hill at the twenty-first milemarker of the Mandalay-Lashio Highway.

Here, a thin spindly Bodhi tree stands alone on the open hilltop surrounded by budding concrete pillars and bamboo awnings. Under its labyrinth of exposed roots and its gnarled branches lie twelve Burmese pythons, spirits entwined to a powerful *thaik* existing within the sacred landscape. The snakes who reside here are said to be kin,[29] brothers and sisters tied through their spiritual ancestry and drawn to one another by the radiant powers of *thaik*.

According to villagers and frequent temple goers, this *thaik* has existed for millennia, but only became known to them in 2014, when the hill was scheduled to be sold for private development along with the other neighbouring hillsides. However, once clearing of the vegetation began under the direction of the township administration, a series of supernatural events occurred and led to the creation of a Snake Temple.

My interviewees recall that when men from the village began removing trees, they reported frequent malfunctions of their clearing equipment, especially when attempting to remove the Bodhi tree. Their mechanistic challenges were reportedly accompanied by vivid dreams, premonitions, in which the village faced a violent demise. The workers became concerned about the project and reported the sources of their apprehension to the township administrator.[30] The administrator interpreted their claims as a sign that the land had sacred qualities or *dago*. He determined that if the construction workers were to continue clearing the land, their ominous dreams would come true, most likely at the hand of resident *nat* spirits. The administrator decided to discontinue the project and declared that the hilltop be reserved for Buddhist worship to improve the karmic fate of local inhabitants. With the help of the community, a small temple was erected in an attempt to appease the guardian spirits. However, signs from the spirit world continued:

> Seven days after the full moon day of Thadingyut (Buddhist New Year), 1376 (10/14/2014), [a couple] found an injured snake near [road, village]. The couple consulted with [the township administrator] regarding the snake's future...
>
> Once [the administrator] and other officials reached the house, [the woman of the couple] became possessed by a *nat* and began talking to [the township administrator], saying that a bad omen would fall upon the entire village if the snake were to be moved elsewhere... [the township administrator] remembered that a new temple compound was being built in the village and wanted to house the snake at the temple.
>
> The next morning (10/16/2014), people from nearby villages visited the temple to see the snake. [A local woman] became possessed by a *nat* spirit, predicting that another snake was on its way and an offering ceremony was to be thrown... Days later, [another local woman] became possessed by a *nat* spirit and said that in 7 days, yet another snake was on her way and asked for another offering ceremony. The possessions proved to be true when two more snakes were found by local villagers.
>
> The snakes were kept on a Bodhi tree growing in the centermost part of the pagoda. [...] On October 20th, 2014, ...under the guidance of [the township administrator], village officials and other locals brought nine buddha images from Mandalay to [village] to establish the Snake Temple.[31]

Since the establishment of the Snake Temple, Burmese pythons have been brought to the temple from places as far away as Yangon by Buddhists who believe that they too share a spiritual ancestry with the *thaik*. Temple goers use the counting word *baa* in reference to the resident snakes, a word which is typically reserved for monks who are considered the most holy members of Buddhist society.

At present, construction of a 108[32] foot zedi, a mound-shaped Buddhist structure, is underway beside the lone Bodhi tree. Dozens of *nat* statues, donations from followers, and over one-hundred Buddha statues from various devotees wait to be housed in the new temple complex. Alms boxes distributed throughout the temple at present are filled with human hair, which will eventually be poured into the cement mixture to build the *zedi*. Those who donate their hair anticipate that they will enjoy prosperous and meritorious future(s) for lifetimes to come, tethered to the holy powers of the Snake Temple.

The story of this Snake Temple, its telling, and its retelling demonstrates the process through which Buddhist cosmologies and *nat* spirits take place. In local narratives, geographical spaces and its biophysical components become entangled with elements of the supernatural realm in the Buddhist imaginary, producing a sacred geography that takes place on the physical landscape.

The land upon which this Snake Temple is now being created was previously fated to be controlled by some private entity, out of the control of the villagers and likely the township administration as well. However, various members of the community took signs from the spirit world, collectively learning that *nat* spirits resided in the landscape. Therefore, it was effectively determined that a Buddhist temple should be built to consecrate the landscape, reserving it for worship by the community.

The recent origins of this Snake Temple illustrate how the future(s) of spirits and their sacred spaces take place within a system of shared knowledge, values, fears and desires of a community of believers. Collective belief in *nat* spirits, also creates new spaces for Buddhism by facilitating social and state recognition of certain landscapes as housing indigenous spirits. Because of the intimate associations between power and Buddhism in the political sphere, designation of novel sacred spaces and the building of Buddhist temples can be made facile by approval of community leaders and township administrators, whose social and religious reputations are bolstered by supporting Buddhist projects.

In the case of Snake Temples, and Buddhist temples in Myanmar more broadly, it is important to emphasize that the building of a temple is not considered as the process by which land is consecrated, or *made* holy. Rather, the land is seen as already having inherent sacred properties which individuals are obliged to respect and cultivate.[33] In the public imagination, the landscapes surrounding Buddhist sites are seen as enlivened by powers that entangle all living and non-living things in a web of spiritual connections. At Snake Temples, this network of relations tethers humans, spirits, and snakes to an existential project of perpetuating the Buddhist religion, taking place upon the geographic landscape. The construction of this Snake Temple produces future(s) for Buddhist ideologies within the compounds of a new sacred site. Simultaneously, new future(s) are produced for *nat* spirits in which they exist as indigenous beings in the land, viewed as integral to the future(s) of people and place.

An economy of merit in Twante Township

On the edge of Twante Township, in Yangon's western outskirts, a Snake Temple is home to over 40 Burmese pythons. The temple, across the river from Myanmar's largest urban centre, is popular among devout Buddhists from surrounding villages and nearby Yangon, as well as a growing number of international tourists. The temple is located along a narrow dirt road between expanses of rice paddies and rubber plantations, where recent land speculation has been attributed for rising property prices and increasing economic insecurity.[34]

According to local residents, the Snake Temple was first constructed during the 1980s,[35] a project which was undertaken by a monk, who was the abbot of the adjacent monastery.[36] Before his death, he sought to build a monument to demonstrate his monastic virtues and earn merit through leading the construction of a new temple. The monk purchased four acres from a local landowner with available funds from the monastery. The landowner, also eager to earn merit for himself, donated an additional four acres to support the temple building project. Once the monk acquired the land, he began receiving[37] vivid dreams from resident *nat* spirits. In these dreams, a snake would appear, explaining to him that the land which he acquired harbours an ancient and powerful *thaik*. The snake advised the monk that other snakes should be installed at the temple, guardian spirits to protect the sacred landscape. The monk shared his knowledge of the supernatural site to his congregations and construction of the temple began.

As news of the new temple spread, Buddhist followers from local villages and across the river in Yangon began frequenting the temple, fulfilling the monk's prophetic dreams by delivering Burmese pythons which were found in the wild or purchased from the wildlife trade. At present, dozens of Burmese pythons roam through the temple spaces surrounded by shrines to the Naga spirits, overflowing with fruits, flowers, and money.

When I ask followers why they choose to worship at this snake temple, they often say that it is because the site has *dago*, the magical quality that enlivens the sacred landscape. Many followers also say that they, and their families, have an ancestral connection or *thaik-hset* with the temple and its guardian spirits. In worshipping at this temple, they not only expect that they will earn merit, but that they will gain favor with the powerful *nat* spirits. These spirits can help them acquire economic wealth, good health in times of illness, and success in business or education, thereby providing a sense of security for future(s) they desire.

Giving donations to local *nats* is an important way that followers can seek spiritual intervention in acquiring wealth while simultaneously demonstrating traditional displays of Buddhist piety. Through these practices, mundane and material desires are not necessarily in conflict with Buddhist ethical evaluations and instead can be seen as complementary with accepted forms of Buddhist worship. Worshipping at Snake Temples allows followers to accrue merit in order to anticipate positive outcomes in future lifetimes as well as exercise control, or gain a sense of determination, over the material conditions in the near future(s) of their present lifetimes. Simultaneously, devotees' donations give continuation to the sacred landscape and strengthen their relation to the spirits.

Religious discourse surrounding this Snake Temple demonstrates the mutual compatibility between Buddhist virtues and desires to accumulate economic resources which are present throughout other Snake Temples more broadly. The emergence of this temple in the late 20th century, and other Snake Temples in my research, should be contextualized within the economic and political shifts in Myanmar which were occurring at the time. Beginning in the 1980s, rapid globalization, a growing capitalist economy, and frequent shifts in the political system, characterized Myanmar's eventual transition from a military junta towards a semi-democratic government in 2011.[38] Important economic and political disruptions during this time include the collapse of the planned economy under the socialist government of General Ne Win in 1988. That same year, a new military government regime, known as the SLORC/SPDC (State Law and Order Restoration Council/State Peace and Development Council)[39] was installed and enacted policies towards modernization of Myanmar and the construction of the semi-capitalist economy.[40] After 2011, Myanmar experienced a period of 'opening up' under a newly installed semi-democratic government under former military general, President Thein Sein, who introduced new forms of market liberalization, globalization, and democratic ideologies.[41] Within this atmosphere of great uncertainty and existential crisis, Snake Temples emerged as a form of 'prosperity Buddhism,' providing followers with a means to cope with unpredictability through seeking help from the spirits to determine positive outcomes in their future(s).

In the origin story of this particular temple, dreams are interpreted as signs or learning(s) from the spirit world, which catalyzed a series of events, bringing the temple into existence. Although the temple was constructed in recent decades, those who worship there believe that they have a *thaik-hset*; a spiritual connection which spans across their previous lifetimes and dates back millennia. Humans, spirits, and snakes are inextricably tied by these connections to the elusive and powerful objects which are embedded in the landscape. This suggests a perception that the spiritual qualities of the landscape and the particular spirits that reside there are indigenous and specific to the land upon which this novel structure is built.

The economic wealth that *nat* spirits bring is to some extent self-realizing insofar as the areas of land they inhabit are often popular sites for pilgrimage. This Snake Temple has become an important site for religious tourism, attracting congregations from surrounding villages and nearby. Through the creation of a new local economy, residents can benefit from the opportunities for income created by tourism to the Snake Temple, both by local residents and international visitors. The popularity of this Snake Temple in particular may be due to its proximity to Yangon, the most densely populated city and largest economic hub in the country. A small economy of drivers, food sellers, and small shop owners who provide services to devotees and tourists has assembled around the holy site. Though the economic activity generated by this Snake Temple holds hardly a flame to larger, more well-known temple complexes in the city centre, prosperity is felt by those who have created livelihoods in relation to the sacred site. In my conversations with shopkeepers and motorbike drivers who cater to visitors of the Snake Temple, they attributed their personal success to the supernatural powers of the pythons. Prosperity is relative, experienced in different ways depending on what is considered good or desirable.

Spirits, snakes and soldiers in Paleik

Another Snake Temple sits along the plains of the serpentine Myit-nge River in Paleik, a settlement not far from Mandalay. There, three Burmese pythons lay coiled on a gilded figure of the Buddha as hundreds of followers wait in line for a glimpse of the elusive deities.

Along the temple walls, a mural tells the story of the temple's origins, which begins in 1976 when a wandering monk came across the ruins of an ancient Pagan Era[42] temple under cover of a dense, uninhabited forest.[43] In the midst of the decay, he saw three Burmese pythons, coiled around a crumbling statue of the Buddha. The carnivorous beasts peacefully resting on the serene image of the Buddha was an indication of the *dago* of the land upon which the pythons were resting. The monk concluded that the pythons were devout spirits, serving as guardians of the ruined temple, enlivened by an ancient *thaik*. He shared his findings with the wider monastic and lay community, who exalted him for his ability to perceive the supernatural, a demonstration of his worldly knowledge. Soon, construction of the temple began with help from devotees from the local village and wealthy donors. The three pythons discovered by the monk became the first to live at the newly established Snake Temple.

In 1994, a fourth snake was delivered to the temple by a military general[44] and his battalion, who were stationed in Mandalay at the time. According to local history, the military official delivered the snake after having a vivid dream in which a snake appeared, pleading to be brought to the temple so that it may fulfill its connection to the *thaik*. Awaking from the dream, the general opened his front door to find a python coiled before him, a sign that the spirit world had crossed into the earthly realm. Accompanied by his battalion, the general delivered the snake to the temple in Paleik, which was also an important stronghold for the military junta. The patronization of the temple by such powerful authorities was met with much celebration by the Snake Temple community, which was a small temple at the time. Such public displays of piety signal to the Buddhist laity that the nation's leaders are moral persons ruling based on Buddhist authority.[45] As interpreted by Julianne Schober, state-sponsored religiosity evokes an aura of kingship, greatness, and continuity of the Buddhist kingdoms of pre-colonial Burma.[46]

As news spread about the military's auspicious donation to the temple, the site attracted a large congregation of devotees. Buddhist followers from local villages, nearby Mandalay, and as far away as Yangon came to worship at the temple. Two years later in 1996, the Mandalay municipal council built a new road for easier access to the temple from the Mandalay-Yangon highway.

Since the creation of the road, the temple has accrued an impressive list of financial supporters, including important members of the Myanmar military and pop-culture icons of the 1990s and early 2000s. Among these include First Lady Daw Khin Khin Win, the wife of former President, General Thein Sein, who served as the 8th president of Myanmar between 2011 and 2016, and well-known celebrity actor Lwin Moe. Donations by such financial supporters have allowed for expansion of the temple, both in terms of infrastructural development of the temple compound as well as its influential reach within the Buddhist community at a national scale.

The history of the Snake Temple in Paleik demonstrates how intersections between state-military institutions and animistic Buddhist traditions create future(s) for individual political actors and fringe, minor Buddhist temples within a system of Buddhist-political authority.

This Snake Temple, as well as the others in my research, emerged under the rule of Myanmar's government by the SLORC/SPDC (State Law and Order Restoration Council), the consecutive military regimes that ruled Myanmar/ Burma between 1988 and 2011. The SLORC/SPDC military junta seized power in 1988 when Burma was facing a great economic crisis and widespread social unrest. During this time, power was wielded by a few military elites and the government faced accusations of human rights abuses by the international community.[47] Though notorious for its oppressive social and economic policies, the SLORC/SPDC military junta also engaged in a wave of systematic temple refurbishing projects. Temple building by the military, though not explicitly stated as such by the military itself, has been interpreted as an attempt at gaining legitimacy through religious means.[48]

Legitimation of state authority by religious displays of morality in Myanmar is not limited to this period.[49] However, elements of *nat* worship, including shrines, murals, and icons, had been admonished by the preceding military regime under General Ne Win (1962–1988).[50] It has been suggested that the recognition of *nat* worship by authorities of the SLORC/SPDC regime, which took control of the government after General Ne Win, was an opportunistic attempt to gain favour with marginalized rural communities which had lost trust in the military government for its failure to provide economic and social stability.[51] By conducting public displays of support for Buddhist institutions, individual political actors could create future(s) for their own political status through religious means. Support for temples which emphasized *nat* worship by the SLORC/SPDC regime also created a political environment in which these previously admonished spirits could return to the environment. Despite oppression under General Ne Win's regime, the *nat* spirits were reinvigorated through political and financial support by powerful actors.

These forms of support perhaps do not have an explicit purpose of state-making through demonstrating religious devotion, but have the consequence of marrying the authority to govern with religious piety, specifically to novel forms of worship in this case. However, it lays the groundwork for a future in which the close relationship between Buddhism and governance remains, producing complex inequalities for non-Buddhists in the country. These points of connection reveal ways in which state legitimacy in Myanmar are grounded in Buddhist ideologies, with implications for the future(s) of governance, sovereignty, authority, and justice for Buddhists as well as Myanmar's non-Buddhist "others."

Conclusion

Snake Temples attract followers seeking to create their own future(s) in the Buddhist cycle of birth and rebirth. At these sites, humans, snakes, and spirits come together – a grounding point in an environment of uncertainty. The rituals and discourses within these sacred sites shed light on the complex entanglements between pre-Buddhist *nat* spirits, communities of contemporary Buddhist practitioners, and sociopolitical frameworks of contemporary Myanmar. Here, indigenous spirits are decoupled from any particular time or group of people.

Snake Temples demonstrate the diverse meanings and functions that *nat* spirits bear. As Foxeus's work has shown, *nats* exist as protectors, or guardians, of sacred Buddhist treasures, or *thaik*, and guarantors of economic and personal success within the Cult of the Guardians of the Treasure trove. In my research at Snake Temples, I find that they fulfill these roles not only as elusive beings within the landscape, but also physical, tangible, living beings on the landscape. Through the incorporation of Burmese pythons into sacred space, Snake Temples exude a unique ethos of coexistence or learning(s) that take place between humans, non-humans, and more-than-humans. Snake Temples are multispecies sacred spaces that could be said to, as Donna Harraway puts it, "make kin."[52] Based on the idea that humans and spirits share an inextricable bond, thought to have formed eons ago and persisting through cycles of birth and rebirth, relations between humans, spirits, and serpents is "something other/more [*sic*] than entities tied by ancestry or genealogy."[53] These relations are reinforced performatively through religious practice and preserved in memory through the telling and retelling of local histories.

Notions that spirits and people are timelessly tied to these places gives hope to followers that the future(s) they desire can be realized. In relying on *nats* to function as protective agents of the Buddhist religion and prosperity in the mundane world, the existence of the spirits is reproduced in cosmologies, sacred places, and imagined future(s). This recursive network of relations creates a landscape in which indigenous spirits take place in the built environment, challenge the uncertainties of capitalist markets, and negotiate the protection of their housing spaces through the religious politics of contemporary Myanmar. However, my exploration[54] of Snake Temples remains shallow still, as these temples are imbued with multitudes of learning(s) and future(s) that have yet to take place.

Notes

1 Based on interviews and documents from six Snake Temples near Yangon and Mandalay, I conclude that Snake Temples (*Mway Paya*), specifically those related to prosperity Buddhism, were first established in Myanmar during the 1980s. This is consistent with the emergence of other individualistic Buddhist practices in Myanmar, which made their early appearance in the 1970s and 1980s and gained greater traction in the following decades.

Ingrid Jordt, *Burma's Mass Lay Meditation Movement: Buddhism and the Cultural Construction of Power*. Athens: Ohio University Press, 14–55, 2007.

Niklas Foxeus, "Possessed for success: Prosperity Buddhism and the Cult of the Guardians of the Treasure Trove in Upper Burma," *Contemporary Buddhism* 18 (1): 116, 2017.

Niklas Foxeus, "Spirits, Mortal Dread, and Ontological Security: Prosperity and Saving Buddhism in Burma/Myanmar," *Journal of the American Academy of Religion* 86 (4): 1123, 2018.

2 Skidmore's introduction to 'Burma at the turn of the twenty first century' outlines major points of transition in the 1980s and the decades that follow. After over two decades of oppressive military rule, led by General Ne Win since 1962, the demonetization of the Kyat currency in 1987 led to widespread economic devastation. This event spurred a series of student-led pro-democracy protests in 1988, known as the 888 Movement. The protests were suppressed with violent military intervention and the ruling General Ne Win was replaced by a new military council known as the State Law and Order Restoration Council (SLORC). Though Western nations imposed sanctions on Myanmar, the regime focused its efforts towards projects of 'modernization' and 'development,' particularly in Yangon and Mandalay. These efforts were continued by the subsequent ruling military regime, operating under the name State Peace and Development Council (SPDC). Military rule continued until 2011, when Myanmar entered a period of reform, including the creation of a semi-civilian and semi-democratic government, the liberalization of the media, and trade with Western nations after the lifting of sanctions.

Monique Skidmore, 'Introduction' in *Burma at the Turn of the Twenty-first Century*, ed. Monique Skidmore. Honolulu: University of Hawaii Press,, 3–5, 2005.

Marco Bünte and Clara Portela, 'Myanmar: The Beginning of Reforms and the End of Sanctions,'Hamburg: GIGA, 2012.

3 Schober argues that among the Buddhist laity, secular powers were and are viewed as an opposing force to Buddhist morals and ideologies. This mistrust of secular influence stems from the British Colonial era (1825–1948), persisted through military rule (1962–2011), and is still salient at present.

Juliane Schober, 'Belonging in a new Myanmar: Identity, law, and gender in the anthropology of contemporary Buddhism,' *Religion and Society* 8: 167, 2017.

4 The Myanmar government has long engaged in armed conflict with minority ethnic groups within the country, particularly groups which are predominantly Christian and Muslim. In recent years, there are heightened conflicts with the Muslim Rohingya ethnic minority in Western Myanmar. Schober highlights that these conflicts have been cited in Buddhist nationalist discourses in recent years as a major threat to Buddhism. (Ibid, 160–161.)

5 Foxeus cites the abovementioned political transitions, economic shifts, and social dislocations as factors that create existential turmoil in the Buddhist imaginary, specifically in the period following the 888 Movement and continuing to the present moment. He situates the rise of prosperity Buddhism within the cosmological uncertainty created by these trends and events.

Foxeus, 'Spirits, mortal dread, and ontological security,' 1108.

6 Foxeus, 'Possessed for success.'

7 Office of the United Nations High Commissioner for Human Rights (OHCHR), 'Declaration on the rights of Indigenous peoples', 2007. http://www.ohchr.org/EN/Issues/IPeoples/Pages/Declaration.aspx (accessed 11 Feb 2020).

8 Isa F. Gómez, 'The UNDRIP: An increasingly robust legal parameter,' *The International Journal of Human Rights, 23*: 7–21, 2019.

9 Siegfried Wiessner, 'The cultural rights of Indigenous peoples: Achievements and continuing challenges,' *European Journal of International Law, 22*: 121–140. 2011.

10 Ken Coates and Carin Holroyd, 'Indigenous internationalism and the emerging impact of UNDRIP in Aboriginal affairs in Canada,' in *The Internationalization of Indigenous Rights* (2014).

Tania Murray Li, Baviskar Amita, Rob Cramb, Kaushik Ghosh, Rusaslina Idrus, Pauline E. Peters, Nancy Postero, Elizabeth Rata, Irina Wenk, 'Indigeneity, capitalism, and the management of dispossession,' *Current Anthropology* 51: 385, 2010.

11 Michael R. Dunford, 'Indigeneity, ethnopolitics, and Taingyinthar: Myanmar and the global Indigenous peoples' movement,' *Journal of Southeast Asian Studies* 50: 51–67, 2019.

12 For Burmese words, I use a 'standard conventional transcription system' recommended by John Okell, a teacher of Burmese language. John Okell, *A Guide to the Romanization of Burmese*. London: Psychology Press, 2002.

13 Eight 'National races' are officially recognized in Myanmar; Bamar, Shan, Kachin, Kayin, Mon, Arakan, and Kayah. Bamar constitutes the largest ethnic group within the country. Despite legal recognition of certain ethnic minorities within the constitution, non-Bamar ethnicities within and outside of the 'national races' are subject to various forms of institutional and legal discrimination.

Nick Cheesman, 'How in Myanmar "national races" came to surpass citizenship and exclude Rohingya,' *Journal of Contemporary Asia* 47: 461, 2017.

14 Grant Brown, 'The pre-Buddhist religion of the Burmese,' *Folklore* 32: 77–100, 1921.

Melford E. Spiro, *Buddhism and Society: A Great Tradition and Its Burmese Vicissitudes*. Berkeley: University of California Press, 1982; Donald Eugene Smith, *Religion and Politics in Burma*. Princeton: Princeton University Press,, 15, 2015.

15 James L. Cox, *From Primitive to Indigenous: The Academic Study of Indigenous Religions*. Oxford: Routledge, 9–31, 2016; Catherine Bell, 'Extracting the paradigm—Ouch!,' *Method & Theory in the Study of Religion*: 114–124, 2008.

Tomoko Masuzawa, *The Invention of World Religions: Or, How European Universalism was Preserved in the Language of Pluralism*. Chicago: University of Chicago Press, 2005.

Jonathan Z. Smith, 'A matter of class: Taxonomies of religion,' *Harvard Theological Review* 89: 387–403, 1996.

16 Iselin Frydenlund, 'Religious liberty for whom? The Buddhist politics of religious freedom during Myanmar's transition to democracy,' *Nordic Journal of Human Rights* 35: 55–73, 2017.

17 Cox, *From Primitive to Indigenous*, 1.

18 Bjørn Ola Tafjord, 'Indigenous religion(s) as an analytical category,' *Method & Theory in the Study of Religion* 25: 221–243, 2013.

19 Here, I rely on Cox's description of indigenous religions in which 'The single and overriding belief shared amongst Indigenous Religions derives from a kinship-based world-view in which attention is directed towards ancestor spirits as the central figures in religious life and practice.'

Cox, *From Primitive to Indigenous:* 69. I draw upon this particular idea to consider *nat* worship at Snake Temples as indigenous because of the concept of *thaik-hset*, which will be explained in the subsequent section. However, I challenge some of Cox's other ideas such as that his definition should be used as a classifier across religious studies.

Ibid, 169–171.

I also challenge his statement that 'there are as many indigenous religious traditions as there are indigenous peoples,' which perpetuates the ontological coupling of Indigenous religions and Indigenous peoples.

Ibid, 1.

20 Niklas Foxeus, 'Spirits, mortal dread, and ontological security: Prosperity and saving Buddhism in Burma/Myanmar,' *Journal of the American Academy of Religion* 86 (4), 2018. Foxeus, #Possessed for success.'

21 Foxeus, 'Possessed for success,' 115.

22 Ibid, 111.

23 I received ethical clearance to conduct research with human subjects from the Institutional Review Board of the University of Wisconsin-Madison.

24 Nicole Tu-Maung, 'The spirit of a serpent; Seeing Buddhism, environment, and politics through the "Snake Temples" of Myanmar' (M.S. thesis, University of Wisconsin-Madison, 2019).

25 Thomas Nathan Patton, 'Phantasmagorical Buddhism: Dreams and imagination in the creation of Burmese sacred space,' *Religions* 9 (12) (2018).

26 Ibid, 415.

27 I borrow the term 'ongoingness' from Donna Harraway as used throughout her book, *Staying with the Trouble, Making Kin in the Chthulucene*. In this collection of essays, 'ongoingness' is used to describe the ways in which different beings come together, make life, and engage in reflexive processes which are less than utopia but more than mere coexistence.

Donna J. Haraway, *Staying with the Trouble: Making Kin in the Chthulucene* (Durham: Duke University Press, 2016).

28 The Cult of the 37 Nats is a group of nat spirits which are widely venerated by Buddhists in Myanmar and were first officially incorporated into Buddhist in the 11th century by King Anawratha.

Bénédicte Brac de la Perrière, 'Being a spirit medium in contemporary Burma,' *Engaging the Spirit World: Popular Beliefs and Practices in Modern Southeast Asia* 5 (2011): 167. Worship of these *nats* represents a more centralized form of spirit worship, which contrasts with the novel cult described in this chapter.

29 Brac de la Perrière, has written on the pervasiveness of sibling relationships in stories about *nats* and analyzes sibling relationships as an important reason for the popularity and resilience of *nat* worship in Burmese culture.

Bénédicte Brac de la Perrière, 'Sibling relationships in the Nat stories of the Burmese cult to the thirty-seven,' *Recherche en Sciences Humaines sur l'Asie du Sud-Est* 5 (2002).

30 Township administrations are headed by the senior official of the General Administration Department (GAD) of the military-led Ministry of Home Affairs.

Michael Lidauer, 'Towards a new state in Myanmar,' in *Burma/Myanmar – Where Now?*, ed. Mikael Gravers et al. Copenhagen: Nordic Institute of Asian Studies, 79, 2014; Hamish Nixon et al., *State and Region Governments in Myanmar*. Yangon: Myanmar Development Resource Institute-Centre for Economic and Social Development and the Asia Foundation, 2, 2013.

31 Untitled carving on stone monolith. Translation provided by Aung Lin Htet. For the privacy of my research participants and their community, names of individuals and specific locations have been removed from the original text.

32 Note that the height of the zedi is a denomination of the number 9, an auspicious number in Burmese Buddhist cosmology.

33 Mandy Sadan, 'Respected grandfather, Bless this Nissan,' in *Burma at the Turn of the 21st Century*, ed. Monique Skidmore. Honolulu: University of Hawaii Press, 98, 2005.

34 My interviewees frequently cited the high price of land and declining access to farming in areas surrounding the Snake Temple.

Further support for this idea comes from Eben I. Forbes, 'On the frontier of urbanization: Informal settlements in Yangon, Myanmar,' *Independent Journal of Burmese Scholarship* 1: 205, 2016.

35 Villagers and temple-goers cited different years in which initial construction began. However, all interviewees reported that construction began sometime in the 1980s.

36 A temple or *zedi* that is located within or near a monastery is considered territorially distinct from the monastery. The monastery is where ordained monks practice meditation and exercise Buddhist discipline. The temple is where Buddhist laity and members of the general public gather to worship the Buddha and hold festivals.

A.W. Sadler, 'Pagoda and monastery: Reflections on the social morphology of Burmese Buddhism,' *Journal of Asian and African Studies* 5 (4): 288–290, 1970.

37 The word *received* is used here because interviewees discussed dreams as messages given to people by *nat* spirits.

38 Foxeus, 'Spirits, mortal dread, and ontological security,' 1108.

39 The SLORC/SPDC (State Law and Order Restoration Council) era is named after the consecutive military regimes that ruled Myanmar/Burma between 1988 and 2011.

40 Stephen McCarthy, 'Ten years of chaos in Burma: Foreign investment and economic liberalization under the SLORC-SPDC, 1988 to 1998,' *Pacific Affairs* (2000).
41 Marco Bünte and Clara Portela, 'Myanmar: The beginning of reforms and the end of sanctions.' Hamburg: GIGA, 8, 2012.
42 The Pagan Era, or Pagan Kingdom (849–1297) was the first dynasty to unite much of present-day Myanmar under a single sovereignty. This laid the groundwork for the growing dominance of Burmese culture and Theravada Buddhism. In studies of local oral histories, much attention is paid to the Pagan Era as it is glorified as the roots of Burmese Buddhism.

Thant Myint-U, *The Making of Modern Burma*. Cambridge: Cambridge University Press, 8, 2001.
43 Descriptive mural *date unknown*. The monk described in the temple's history is a *forest monk*, or one that enters into the forest for meditation. A monk's association with the forest signifies a superior degree of spiritual achievement. He is considered learned in the spiritual world. For more on the spiritual knowledge(s) and learning(s) of forest monks in Myanmar, see Guillaume Rozenberg, *Renunciation and Power: The Quest for Sainthood in Contemporary Burma*. New Haven: Yale University Southeast Asia Studies, 21, 2010.
44 The identity of the general is likely known to the temple committee (*gawpaka*), an appointed group of respected male elders, but has not been made known to me. However, visual evidence such as photographs support evidence of military presence at this particular temple.
45 Jordt, *Burma's Mass Lay Meditation Movement,* 194–195.
46 Ibid, 188–190.
47 Michael W. Charney, *A History of Modern Burma*. Cambridge: Cambridge University Press, 147, 2009.
48 Janette Philp and David Mercer, 'Commodification of Buddhism in contemporary Burma," *Annals of Tourism Research* 26, 1999.

Jordt, 'Burma's mass lay meditation movement,' 188–199.
49 Philp and Mercer, 'Commodification of Buddhism,' 39–40.
50 Donald Eugene Smith, *Religion and Politics in Burma*. Princeton: Princeton University Press, 350, 1965.
51 Yves Rodrigue, *Nat-Pwe: Burma's Supernatural Sub-Culture*. Gartmore, Scotland: Kiscadale Publications, 1992.

Bruce Matthews, 'The present fortune of tradition-bound authoritarianism in Myanmar,' *Pacific Affairs* 71 (1), 1998.
52 Haraway, *Staying with the Trouble,* 102–103.
53 Ibid, 102–103.
54 This research would not have been possible without the monks and lay practitioners who allowed me to learn from their sacred spaces and precious teachings. I am indebted to Aung Lin Htet for acting as a research partner and friend throughout interviews and beyond. I thank Dr Timothy Van Deelen, Dr Paul Robbins, and Dr Tun Myint for providing academic and intellectual support for my M.S. thesis on this subject. I am grateful for the support of Matthew Venker, with whom I discussed many of the ideas in this chapter before putting them into writing.

References

Bell, Catherine. 'Extracting the paradigm—Ouch!' In *Method & Theory in the Study of Religion*, 114–124, 2008.

Brac de la Perrière, Bénédicte. 'Sibling relationships in the Nat stories of the Burmese cult to the thirty-seven.' *Recherche en Sciences Humaines sur l'Asie du Sud-Est* 5: 31–48, 2002.

Brac de la Perrière, Bénédicte, 'Being a spirit medium in contemporary Burma,' *Engaging the Spirit World: Popular Beliefs and Practices in Modern Southeast Asia* 5: 163–183, 2011.

Bünte, Marco, and Clara Portela. 'Myanmar: The beginning of reforms and the end of sanctions.' Hamburg: GIGA, 2012.

Charney, Michael W. *A History of Modern Burma.* Cambridge: Cambridge University Press, 2009.

Cheesman, Nick. 'How in Myanmar "National Races" came to surpass citizenship and exclude Rohingya.' *Journal of Contemporary Asia* 47: 461, 2017.

Coates, Ken, and Carin Holroyd. 'Indigenous internationalism and the emerging impact of UNDRIP in Aboriginal affairs in Canada.' In *The Internationalization of Indigenous Rights*, 2014.

Cox, James L., *From Primitive to Indigenous: The Academic Study of Indigenous Religions.* Oxford: Routledge, 2016.

Dunford, Michael R. 'Indigeneity, ethnopolitics, and Taingyinthar: Myanmar and the Global Indigenous peoples' movement.' *Journal of Southeast Asian Studies* 50: 51–67, 2019.

Eugene Smith, Donald. *Religion and Politics in Burma.* Princeton, NJ: Princeton University Press, 1965.

Forbes, Eben I. 'On the frontier of urbanization: informal settlements in Yangon, Myanmar.' *Independent Journal of Burmese Scholarship* 1: 197–238, 2016.

Foxeus, Niklas. 'Possessed for success: Prosperity Buddhism and the cult of the guardians of the treasure trove in Upper Burma.' *Contemporary Buddhism* 18 (*1*): 108–139, 2017.

Foxeus, Niklas. 'Spirits, mortal dread, and ontological security: Prosperity and saving Buddhism in Burma/Myanmar.' *Journal of the American Academy of Religion* 86 (4): 1107–1147, 2018.

Frydenlund, Iselin. 'Religious liberty for whom? The Buddhist politics of religious freedom during Myanmar's transition to democracy.' *Nordic Journal of Human Rights*35: 55–73, 2017.

Gómez Isa, F. 'An increasingly robust legal parameter.' *The International Journal of Human Rights*, 23: 7–21, 2019.

Grant Brown, R. 'The Pre-Buddhist religion of the Burmese.' *Folklore* 32: 77–100, 1921.

Haraway, Donna J. *Staying with the Trouble: Making Kin in the Chthulucene.* Durham: Duke University Press, 2016.

Jordt, Ingrid. *Burma's Mass Lay Meditation Movement: Buddhism and the Cultural Construction of Power.* Athens, OH: Ohio University Press, 2007.

Li, Tania Murray, Baviskar Amita, Rob Cramb, Kaushik Ghosh, Rusaslina Idrus, Pauline E. Peters, Nancy Postero, Elizabeth Rata, Irina Wenk. 'Indigeneity, Capitalism, and the Management of Dispossession.' *Current Anthropology* 51: 385–414, 2010.

Lidauer, Michael. 'Towards a New State in Myanmar.' *Burma/Myanmar – Where Now?*, edited by Mikael Gravers and Gravers Ytzen, 72–86. Copenhagen: Nordic Institute of Asian Studies, 2014.

Masuzawa, Tomoko. *The Invention of World Religions: Or, How European Universalism was Preserved in the Language of Pluralism.* Chicago, IL: University of Chicago Press, 2005.

Matthews, Bruce. 'The present fortune of tradition-bound authoritarianism in Myanmar.' *Pacific Affairs* 71 (1): 7–23, 1998.

McCarthy, Stephen. 'Ten years of chaos in Burma: Foreign investment and economic liberalization under the SLORC-SPDC, 1988 to 1998.' *Pacific Affairs*: 233–262, 2000.

Myint-U, Thant. *The Making of Modern Burma*. Cambridge: Cambridge University Press, 2001.

Office of the United Nations High Commissioner for Human Rights (OHCHR). 'Declaration on the rights of Indigenous peoples', 2007. http://www.ohchr.org/EN/Issues/IPeoples/Pages/Declaration.aspx (accessed 11 Feb 2020).

Philp, Janette, and David Mercer. 'Commodification of Buddhism in contemporary Burma.' *Annals of Tourism Research* 26: 21–54, 1999.

Okell, John. *A Guide to the Romanization of Burmese*. London: Psychology Press, 2002.

Patton, Thomas Nathan. 'Phantasmagorical Buddhism: Dreams and imagination in the creation of Burmese sacred space.' *Religions* 9 (12): 414–428, 2018.

Rodrigue, Yves. *Nat-Pwe: Burma's Supernatural Sub-Culture*. Gartmore, Scotland: Kiscadale Publications, 1992.

Rozenberg, Guillaume. *Renunciation and Power: The Quest for Sainthood in Contemporary Burma*. New Haven, CT: Yale University Southeast Asia Studies, 2010.

Sadan, Mandy. 'Respected grandfather, bless this Nissan.' In *Burma at the Turn of the 21st Century*. Monique Skidmore, ed. Honolulu, HI: University of Hawaii Press, 90–111, 2005.

Sadler, A.W. 'Pagoda and monastery: Reflections on the social morphology of Burmese Buddhism.' *Journal of Asian and African Studies* 5 (4): 282–293, 1970.

Schober, Juliane. 'Belonging in a new Myanmar: Identity, law, and gender in the anthropology of contemporary Buddhism.' *Religion and Society* 8 (1): 158–172, 2017.

Skidmore, Monique. 'Introduction.' In *Burma at the Turn of the Twenty-first Century*. Monique Skidmore, ed. Honolulu, HI: University of Hawaii Press, 1–17, 2005.

Smith, Jonathan Z. 'A matter of class: Taxonomies of religion.' *Harvard Theological Review* 89: 387–403, 1996.

Spiro, Melford E. *Buddhism and Society: A Great Tradition and Its Burmese Vicissitudes*. Berkeley, CA: University of California Press, 1982.

Tafjord, Bjørn Ola. 'Indigenous religion(s) as an analytical category.' *Method & Theory in the Study of Religion* 25: 221–243, 2013.

Tu-Maung, Nicole. 'The spirit of a serpent; seeing Buddhism, environment, and politics through the "Snake Temples" of Myanmar.' M.S. thesis, University of Wisconsin-Madison, 2019.

Wiessner, Siegfried. 'The cultural rights of Indigenous peoples: Achievements and continuing challenges.' *European Journal of International Law*, 22: 121–140, 2011.

5 Dreaming of the future(s)

An exploration of the dreams and resistance of the Obo-Manobo

Grace Simbulan

Mt. Apo, an integral site of Obo-Manobo resistance, has constantly been embroiled in conflict amongst private corporations, and various governments who have encroached on Obo-Manobo land threatening their pasts, presents and future(s). Bo-i was one of the few who opposed bloodshed and resisted through the establishment of the cultural regeneration movement. A key part of this movement was the reincorporation of dreams and their interpretation as a part of Obo-Manobo community life. All these efforts united the community and solidified the struggle against infiltration and exploitation by external forces. Bo-i and her clan have used dreams to bring the presence and knowledge of their ancestors to their decision-making process to guide and protect their clan and its future(s).

This chapter explores the contemporary role of dreams in the community life of one of the Lumad[1] groups in Mindanao, called the Obo-Manobo. I draw on the reflections and personal narrative of a female leader of the Obo-Manobo commonly known by her honorary title of *bo-i* (pronounced BAH-ee). This title reflects her authority as a social and spiritual leader, dreamer, and interpreter of dreams. I argue that dreams are a crucial form of learning for resistance for Obo-Manobo leaders in so far as they bring to the present the knowledge of the ancestors to ensure the continuation of life of the Obo-Manobo people-and-land. In this chapter, I draw on Gerald Vizenor's survivance[2] to conceptualize resistance and explore the role dreams play in it.

I have listened, and I now retell the stories Bo-i has allowed me to share. Retelling requires yet another layer of translation and interpretation. The interpretation I present here narrates Bo-i's story as a kind of dream to maintain the quality of Bo-i's own recollections and the survivance of her people-and-land, which she shared with me last year (2019). The focus is on the role of dreams as resistance in Bo-i's own life's journey. The chapter is divided into six sections. The first section situates Bo-i's story through a geographical and historical description of the Obo-Manobo people and their sacred ground – Mt. Apo. The second section expounds on Bo-i's role as a dreamer within her community. This also covers ethnographic accounts by previous Filipino scholars who engaged with the Obo-Manobo of Mt. Apo, and other Lumad groups in Mindanao. The third section talks about how Bo-i's childhood away from

Mt. Apo influenced her future resistance. The fourth section describes how Bo-i's dreams guided her return to Mt. Apo after being away for 25 years. A detailed account is included in the fifth section of how Bo-i's dreams were able to guide her to establish the cultural regeneration movement. The final section describes Bo-i's role as a leader in exile.

I present these sections as a testament to the (hi)story and culture that Bo-i has described, a memory of the resistance of her people. My aim is to present a decolonized perspective on Bo-i's (hi)story by exploring the ways her clan records and transmits memory and knowledge through Bo-i's dreams and its interpretation. In this space, I encourage readers to see dreams and dream-sharing as a critical means of documenting and passing on Indigenous knowledge as a means of securing their future and survivance as Obo-Manobo people-and-land.

Defining space, taking place

> Nakatira noon sa Balabag ang lolo at lola kong si Omme pero araw-araw siyang nagpupunta sa Inanapo. Tinawag siyang *anituwon* kasi nakikipag-usap siya sa mga espiritu. Araw-araw nagpupunta sa gubat ang lola ko kaya nagseselos ang lolo ko. Madalas, hapon na siyang umuuwi at may dala-dalang mga pagkaing-gubat. Sabi sa kwento, isang araw yung lola ko hindi na bumalik sa lolo ko. Tapos natagpuan siyang patay doon sa Inanapo.
>
> My grandfather lived in Balabag with Omme, my great grandmother, but every day she would go to the forest, to Inanapo. The people called her an *anituwon*, someone who talked to the spirits of the forest. She did this so often that even my great grandfather became jealous. Every day, she'd return home in the afternoon, carrying food from the forest. But one day, according to the legend, my great grandmother never returned. After a long search, Omme's lifeless body was found, there, in the forest of Inanapo.

I open Bo-i's story with a vignette of her great grandmother's life story, Omme, because of two reasons. First Omme's story was often told to Bo-i as a child and the dreams that defined Omme's life and her interaction with it laid the foundation for Bo-i's own story. Second, this story is a testament to how Bo-i's clan envisioned the world. They believed that dreams connected them to the spiritual world and to their land, which here I introduce to you through Bo-i's narrations.

Bo-i believes that her ability to interact with dreams and spirits was passed on to her by her great grandmother Omme, and Bo-i shared Omme's vision of the world – centered around Mt. Apo, the highest peak in the southern Philippines. "Naniniwala kami na may dalawang mundo— yung mundo pag gising tayo at yung pag natutulog tayo. Kapag tulog kasi tayo, unguarded tayo kaya mas madali kaming makausap ng mga ancestors namin. (We believe that humans have two worlds – the waking life, and the world when we're sleeping. When asleep, we are unguarded, so it's easier for us to talk to our ancestors.)" She also described her ancestors' belief in the afterlife and how Mt. Apo is perceived as an extension of themselves as people-and-land. "May kabilang-buhay kaya ang kaluluwa namin ay babalik sa Mt. Apo, (There's an after-life and our souls will return to Mt. Apo)," she said. "Sa Mt. Apo naninirahan ang aming mga kaluluwa kaya itinuturing namin itong sagrado. (Our souls reside in Mt. Apo, so this is the reason why we consider it sacred ground.)"

The geography of Bo-i's childhood bore little resemblance to the political maps that one sees today, instead relying on her people's understanding of the world around them. "Kapag nakaharap ka sa Mt. Apo (When facing Mt. Apo)," she said,

> "Nasa kaliwa yung Pandanon, kung nasaan yung kagubatan. Mowoo yung ilog na nasa pagitan ng Pandanon at Sayaban. Sa kanan naman ng Sayaban nandoon yung Matingaw River na namamagitan sa Sayaban at Balabag. Balabag yung lupaing minana ng lolo ko."
>
> On the left side is Pandanon and it's where the forest is. Mowoo is the river separating Pandanon from Sayaban. To the right of Sayaban is Matingaw River dividing Sayaban and Balabag. Balabag is the land my great grandfather inherited.

When the borders were set by the national government, Inanapo, which once was not included in any political boundary, was integrated into Sayaban. Bo-i and her clan understood and identified with their environment through river systems and various natural landmarks. This consciousness over space resonates with Lumad oral traditions, in which social ties among the Lumads were organized according to river systems as opposed to the present-day cultural and political divide between the upland and lowland communities now dominated by mainstream city settlers.[3]

The tension between this way of seeing and the way of people from outside Mt. Apo has led to conflicts over generations. Memories of resistance against encroaching interests by private corporations, and the Spanish, American, and Philippine governments continue to penetrate the consciousness of the Lumad people. From the early 17th until the late 19th century, the Lumad people resisted all attempts at religious conversion by Spanish Christian missionaries and military troops.[4] During the American colonial period, a change occurred which had a significant impact on the lives of many Manobos. The Philippine Constabulary and American officers put an end to intertribal wars among the Manobos. The civil government, on the other hand, encouraged the datus (Manobo Indigenous leaders), to leave the hinterlands and reside in villages where schools were much more accessible.[5] Since the 1950s, many non-Lumads from the nearby cities have occupied and settled permanently in Indigenous territory, forcing many Manobos to give up portions of their land.[6] Despite these pressures, Mt. Apo is considered an integral site, ecologically and politically, by several ethnolinguistic groups in the area, including the Jangans, Atas, Tagakaolo, and the Manobos in Cotabato[7] who believe that their ancestors and their mountain god, Apo Sandawa, live in the mountain, guiding and protecting the Obo-Manobo people-and-land through their dreams and its sharing and interpretation.

In what follows, I describe the role of dreams in Bo-i's clan and that of the Obo-Manobo more generally. In doing so, I also describe Bo-i's earliest experiences with dreams, and how she came to be known as a dreamer. Bo-i's close connection to these ideas during her childhood had a profound impact on her decisions throughout her adult life. Omme's passing in Inanapo, and Bo-i's eventual return to Mt. Apo after twenty-five years of being uprooted, has led me to believe that Obo-Manobo's strong connection with Mt. Apo has eventually led them back to their land.

Dream-literacy in interpreting future(s)

> We who dare must lick our wounds
> Turn to Magbabaya to lift our thorns
> We may have tilt on the losing side
> But on his grace we can depend
> We who dare are his seeds on earth
> We who dare must keep moving on
> Amidst strong surge and storms
> Our domains we need to protect
> And to Monama we raise our case
> He will lead us to our dream and rest.
> (Excerpt from the poem "We Who Dare" by Bo-i)

An important resistance strategy of the Obo-Manobo is reviving their communal knowledge and dream interpretation as a way of knowing. I refer to dream interpretation as communal knowledge because dreams are explicitly from and of the Obo-Manobo community. In Obo-Manobo culture, when someone is visited or gifted with knowledge through dreams, the dreamer must share the dream to those who will be directly affected by the dream. The dreams gain meaning through attention, and they gain meaning for the givers of that attention through careful communal study. The dreams can be described as a product of the community, even though they in turn create the community from which they came.

Although the precise meaning and function of dreams within Obo-Manobo culture resists easy definition, dreams are an important aspect of Obo-Manobo cultural legacy. Bo-i considers this ambiguity and ambivalence a core feature of the dreams' role in her clan's resistance. The very definition of "resistance" has been so coloured by colonial connotations that I conceptualize resistance by drawing on Anishinaabe scholar, Gerald Vizenor, who defines survivance as "an active sense of presence, the continuance of native stories, not a mere reaction, or a survivable name."[8] Stories of survivance are therefore "renunciations of dominance, tragedy and victimry."[9] Dreaming and interpreting dreams is a form of resistance because it fosters the continuation of Obo-Manobo's ways of knowing and anticipating the future of their people-and-land. In response to colonization by external forces which threaten Obo-Manobo's pasts, presents and futures, Bo-i and her clan have used dreams and their interpretation to bring the presence and knowledge of their ancestors to their decision-making process to guide the clan. Moreover, dreamers and interpreters weave stories of pasts, presents and futures that inform other non-confrontational resistance actions linked to culture, as will be explained below. At the centre of Bo-i's story as presented here, Bo-i's own life story, poems, and words are a dreamed resistance that refutes attempts to be silenced in exile away from the Philippines.

The belief in the guiding role of dreams is also shared by other Lumad peoples, including the Mandaya of Davao del Norte and the T'boli of Lake Sebu. Parong Alimbon, a Mandaya, said that balyans (spiritual leaders) in their culture are able to cure the sick by combining panawagtawag (prayer) and herbal medicines. They believe that the spirits and Magbabaya (Almighty) who dwell in the rivers, mountains and forests, provide them with knowledge to cure, through dreams. The T'boli people believe that the spirits that dwell in Lake Sebu and the surrounding mountains live among the community members. They have been passing on their Indigenous knowledge through a woven material known as T'nalak. Just like dream-sharing in Bo-i's clan, T'nalak patterns are learned through dreams, which they believe are gifts from Fu Dalu (the spirit of hemp). As a result, the T'nalak weavers are known as "dream weavers".[10] For example, the ability to dream and learn T'nalak patterns are passed on matrilineally. Inas Cone of Lemkadi, a dream weaver, described a T'nalak pattern she learned from her mother, who learned it from Cone's grandmother. The pattern depicted a bwengkel (crocodile). According to T'boli oral history, an illicit relationship between two relatives caused the spirits of the lake to curse and transform them into crocodiles as a form of punishment. The T'boli people believe that the couple still inhabit the lake to this day. This pattern, which was passed on to Cone's ancestors through dreams, communicates Indigenous wisdom that has survived for generations[11]. Cone and other dream weavers play a crucial role in the passing on of Indigenous knowledge through their dreams and its interpretation. The T'boli people believe that knowledge is passed on to them and blessed by Fu Dalu to protect, warn, and guide them from danger as a means of their survivance as T'boli people-and-land.[12]

Given the importance ascribed to dreams in Lumad culture, it was not a surprise to learn that community dream-sharing and interpretation has been considered an important practice of Bo-i's clan dating back many generations. According to Bo-i, dreams in their clan function primarily as an early-warning system, protecting them against danger. "Sa amin, (In our clan)" she said in a mix of English and Tagalog, "ang dreams ay for protection at warning para sa isang komunidad hindi para sa sarili lang kaya kami nagsheshare ng dreams. (The community uses dreams for their protection and warning against danger. This is why I share dreams.)" Bo-i described the use of dreams in this way as a long-standing practice among the Obo-Manobo. For example, Bo-i said that the dreams of pre-colonial Obo-Manobos in Mt. Apo predicted the arrival of the American colonists at the turn of the 20th century. According to the story, dreamers in the community dreamt of coming face to face with giants. In their dream, they heard booming sounds coming from Mt. Apo resembling a volcanic eruption. Later, as a group, the Obo-Manobo interpreted this as the sound created by the weapons of the Americans. They interpreted these dreams to mean that they should avoid confrontation with the Americans by hiding in the mountains. This knowledge was passed on to each generation through oral tradition. In this way, the passing of knowledge through dreams plays an important role in bridging their past, present, and future(s). Bo-i believes that dreams are part of their Indigenous identity as it plays a dialogical role that traverses physical and spiritual boundaries. Dreams provide a dream-space where everything takes place. This spatial location is where Obo-Manobo ancestors, the spirits of Mt. Apo, the souls of the departed, and the dreamer all participate in the passing and retelling of Indigenous knowledge to guide, protect, and ensure the continuation of life and survivance of the Obo-Manobo people-and-land.

In Bo-i's experience, dreams' meaning is defined through the interpretation of the dreamer and the additional discussion and interpretation with the community. I refer to Bo-i's ability to provide sound dream interpretations as "dream-literacy" which she developed and honed through years of observing and participating in dream-interpretations with members of her clan. The solution to those problems is the dream's meaning. By "solving" the dream, the Obo-Manobo teach and learn together, simultaneously strengthening the bonds of the community. When Bo-i was a child, dream interpretation was an important ritual every morning. It often took precedence over other activities as described in the following section.

Bo-i's calling as an interpreter of dreams

In Bo-i's experience, dreams were typically shared through oral description. Bo-i herself sees this communal aspect as an important difference between the Obo-Manobo people-and-land and many outsiders.

> May mga tao talagang dreamer tapos sa aming katutubo dahil pinapahalagahan namin kaya siya functional. Sa Western world si Sigmund Freud lahat ng interpretation all about the self. Hindi ako naniniwala kay Freud, kasi ang kanyang definition of dreams para sa sarili. Eh ang aming definition of dreams ay para sa lahat, meant to be shared tulad ng warning, may darating na gutom, tulad ng lola ko may darating, na pestilence [at naiwasan niya ito].
>
> Dreams are functional to us Indigenous peoples because we take them seriously. In the Western world, Sigmund Freud's interpretation of dreams is all about the self, but our definition of dreams is for everyone. It's meant

> to be shared like warnings such as the imminence of famine. An example of this is when my great grandmother dreamt of pestilence [and was able to prevent it].

Bo-i presents dreams as active and alive. For Bo-i, there is an active and conscious participation and conversation coming from the dreams and the dreamer. Not everyone in Bo-i's Indigenous community dreamt or was able to remember their dreams vividly. However, when someone did, the community leaders and elders – referred to as the *boyahhon* – often interpreted the dream and ascribed meaning to certain dreams that were presented. In every clan, the appointed dreamers were included in the *boyahhon* to help in decision-making. They considered someone a dreamer if her dreams were considered valuable by her community's *boyahhon*. Ancestors and spirits revealed themselves through dreams and those who experienced them were chosen based on their good character. In Indigenous communities in Mindanao, the intermediaries between the spirits and the community were sometimes known as babaylan/baylan/balyan/waylan/mabalian/beliyan/magbulungay[13]. Babaylans, who were predominantly women[14], exercised social and spiritual power in pre-colonial Philippines. The traditional path in babaylan formation was to be called by ancestors through dreams[15]. They have "the ability to mediate with the spirit world, has her own spirit guides, and is given gifts of healing, foretelling, and insight. She may also have knowledge of healing therapies such as *hilot, arbularyo*. She has the gift of travelling to the spirit world or non-ordinary states of reality in order to mediate with the spirits."[16] According to Bo-i, babaylans in their clan are also considered as the dreamers who also participated in dream interpretation for early warning. For this reason, dreams play an important role during the hunting season. The night before the men hunt, warnings come from the land and ancestors to the Obo-Manobo in the form of dreams. For example, Bo-i described how it came to be known that if anyone in the community dreamt of running after three humans, it meant the spirits have only allowed them to take three deer or other animals. Taking more than what was allowed resulted in punishment by the spirits of Mt. Apo through sickness or an unfavourable event. Bo-i also mentioned that her ancestors were able to find a cure for certain diseases and avoid being poisoned by deadly plants in the forest by dreaming about the antidote or herbs they would need.

In Bo-i's case, her role as dreamer – as an intermediary between the waking world and the spirit world, manifested at the early age of five. There were no prescribed qualifications for dreamers or dream interpreters, nor were there prescribed methods for interpretation. This early discovery of Bo-i's gift was in part due to Bo-i's participation in community decision-making at a very young age, when she would accompany her parents during their work as leaders and conflict mediators in their community. Because Bo-i's father frequently worked outside the Indigenous community, Bo-i became more attached to her mother. This physical separation from her father, she says, likely contributed to making her feel closer, spiritually, to her mother's way of resolving disputes.

Her mother's loyalty to serve and ability to resolve disputes encouraged many members of their clan to seek her help to resolve disputes in the community. As a result, her mother became an honorary member of their clan's *boyyahon*. Bo-i recounted nightly meetings in their house where she and her sister experienced the *boyyahon*'s long discussions:

> Kung ano yung mga pinaggagawa ng nanay ko andun ako. Example may isesettle na conflict. Makikinig lang ako hanggang makatulog na ako sa sahig kasi wala namang bangko bangko. Kasi sa tribo (hindi) pinapatulugan ang mga gulo. Yung mga pag-uusap ng mga matatanda hindi yan matatapos hangga't meron silang solusyon na makikita. Kung may penalty, kung may kailangan ipalayo– lahat meron ng solusyon yan. Kaya ako madalas kong naririnig yung pag-uusap ng matatanda, yung pagdedebate, may mga pros and cons tapos yung nanay ko may say din siya doon kahit siya pinakabata, pinapakinggan din siya. Kasi yung nanay ko, I remember kasi eldest siya masyado siyang assertive.
>
> I was always with my mother wherever she went. For example, she'd settle a conflict somewhere. I'd listen to them until I fell asleep on the floor because in our tribe, we never sat around a table. In our tribe, we never slept until disputes were resolved. Adults would have endless conversations until they found a solution. If there's a needed penalty, if someone needs to be sent away—everything has a solution. I always listened to what the adults talked about, what they debated on, the pros and cons. My mother had a say on things even if she was the youngest [in the group]. They listened to her. As the eldest in her family, she was very assertive.

Because these sessions often took many hours to finish, Bo-i frequently fell asleep at her mother's side. When Bo-i awoke, she would describe what she saw in her sleep to the Obo-Manobo leaders and elders. She often dreamt of her people-and-land. "Parang nakita yun ng matatanda na ako'y may tinatawag na *dowon od tanod*, (The elders saw that I was called by the *dowon od tanod*)," Bo-i said, "na ang ibig sabihin ay mayroong gabay. Parang may guidance in different forms – ancestors and spirits (which means that I am guided by different forms, like the ancestors and spirits)."

This period of childhood observation and immersion ended abruptly following the death of Bo-i's mother when she was seven years old. In our conversations, Bo-i recalled having dreams about her mother for three consecutive nights before her mother passed away. In the dream, she saw the ground opening up and swallowing her mother. In one separate dream, she heard her mother's voice from under the earth while she was walking towards the forest. At that time, Bo-i did not know how to interpret the dreams, and she did not share them with the community. Looking back, Bo-i said that her dreams were in fact a warning to her and her family.

In Bo-i's experience, dreams in Mt. Apo played a vivid role in her childhood and had a profound impact on her professional and personal decisions for the rest of her life. Today, dreams are an integral part of Bo-i's work as an activist on behalf of her people-and-land, but when Bo-i was forced to leave her Obo-Manobo community at the age of seven, she was strongly encouraged to forget her Indigenous identity. In the following section, I will describe Bo-i's journey from the foot of Mt. Apo to Kidapawan City, and the period of forgetting that this move entailed. More than two decades later, this was followed by a renewed concern for and interest in working on behalf of her people-and-land. This in turn led to a renewed interest in her dream-literacy and role as a dreamer.

Dreamer away from dream-space

From 1970 until 1995, Bo-i and her two siblings lived with her father in Kidapawan City, five or so miles from their community at the foot of Mt. Apo.[17] Bo-i's father was a Waray from the island of Samar, 300 miles away from Mt. Apo. During this period (1970–1995), as Bo-i grew from a child into a young adult, she was encouraged to ignore her dream-ability and dream-literacy. As a result, Bo-i experienced a period of spiritual upheaval and alienation.

Between her arrival in Kidapawan at age seven and her enrollment in high school at age 13, Bo-i was not allowed to return to Mt. Apo. She vividly remembers when martial law was declared in 1972. She was in grade five. "Doon ko naintindihan na yung mundo pala may ibang namumuno– meron palang nakasulat na batas sa labas ng lupaing ninuno (This was the time I realized that there's another form of leadership [referring to political system in the context of the conversation] that ruled people outside the Indigenous community)," she said. "Noong elementary ako, inintroduce yung constitution, tapos pinamemorize yung Bill of Rights. (In elementary, we were introduced to the Philippine Constitution and we had to memorize the Bill of Rights.)" She said it was the first time that she felt she actually understood the concept of government – what it was, and what it aimed to do.

Bo-i was also aware of political divisions within her father's family, although she says she did not appreciate the full context of this conflict. Her father's support for former Philippine dictator Ferdinand Marcos caused a huge rift between her father and some of her siblings. Back then, Bo-i's father worked as barangay captain (head of the village) for close to 40 years, under five national administrations until Marcos' downfall in 1986. Bo-i's nine years' experience working in a human rights law office, in addition to her living in Kidapawan City at the beginning of martial law, exposed her to many human rights abuses committed by the Marcos administration. Bo-i knew that her siblings also actively opposed the Marcoses.

Bo-i says that during this time of political upheaval, she also felt significant spiritual confusion, much of which was caused by members of her own family. "Ang tingin ng mga brothers ko yung Indigenous, uneducated, ignorant (My brothers considered Indigenous peoples as uneducated, backward, and ignorant)." She said:

> Kaya pilit akong ginawang hindi Indigenous ng mga kapatid ko kasi ayaw nila kaming madiscriminate, para mataas yung tingin sa amin ng lipunan

kailangan talikuran namin yung pagiging Indigenous namin. Sa mga kapatid ko yung pag-cha-change nila sa amin ay para sa aming kabutihan but di nila narealize na yung pilit na pagdisconnect namin ay makacause ng confusion or internal conflict sa loob namin (at) di nila naisip yun.

They forced me to not become Indigenous [referring to conformity to city behaviour and rules in the context of our conversation]. I knew that they meant well in their efforts to change us because they didn't want us to be discriminated against. Little did they know that the identity confusion, and internal conflict we experienced in the city were due in part by my stepbrothers' efforts to disconnect from our Indigenous roots.

Bo-i specifically remembers confusion over the new social rules she was asked to follow.

Assignment ko palagi after meals, paputiin yung gilid ng kaldero. Walang kaldero na maitim na nasa labas kasi nakakahiya daw sa bisita. Marami naman kaming bisita sa bundok [referring to Mt. Apo] hindi naman sinabi sa nanay ko na dapat kaming mahiya sa bisita. So yung idea na kailangan iplease mo sa mata whoever is the visitor kailangan maging pleasant yung, yung mga idea na ganyan kailangan yung paligid palaging perfect, pleasant to the eyes of the visitors.

My task after each meal was to scrub the bottom of our pots and pans to turn it back to its original color. I was asked to scrub all the metalware because they were ashamed to invite guests over. In Mt. Apo, we always had guests over but my mother never told us to do these things in order to please our guests. This idea of always pleasing the guests and the idea of having a perfect environment so it'll be pleasant to the eyes of the visitors [were hammered into us].

While Bo-i and her siblings were growing up in Kidapawan, Bo-i helped her nieces and nephews with their homework. She remembers in particular helping them to write short essays about the American War in Vietnam. She considered herself an introvert at that time, but found that she enjoyed writing, which gave her a feeling of personal release. Bo-i's love of writing has stayed with her ever since. The poems that punctuate this chapter are the latest iteration of her life-long writing practice.

Dreams as calls to return to Mt. Apo

We always tilt on the losing end
Where death is just a step behind
We dare to take every challenge
When knocked by the power of change
We ask for nothing but life.

(Excerpt from the poem "Who We Dare" by Bo-i)

By the end of the 1980s, Bo-i began a long path that would lead her ultimately to serve her community. In 1988 in Cotabato, Bo-i was introduced by her sister to political officers of a national women's leftist group. When Bo-i first saw her sister with other members of the organization, Bo-i thought they were just a group of friends. She didn't pay much attention. After seeing them meet several times in the house, however, and overhearing some of their conversations, Bo-i realized that they were activists who fought against the Marcos administration and were then continuing the fight against oppression perpetuated by the Aquino administration. At the time, the organization focussed on documenting the negative impacts of the ongoing conflict in Mindanao between the Philippine Military and the armed wing of the Communist Party of the Philippines, on Indigenous women and children. Bo-i's introduction to the women's organization was her first introduction to activism in the city.

Because all the organization's meetings were held in the family home, her sister and her colleagues gradually got to know Bo-i and saw her talent in writing through samples Bo-i showed them. They decided to invite her to document the human rights abuses committed by the Aquino administration in Mindanao. Bo-i accepted the challenge. The work filled her with hatred against the perpetrators of what she and her colleagues perceived as injustice. As a result of this work, she also became more familiar with different factions of the Philippine Left, which included some groups that advocated armed rebellion in order to stop the injustices. Bo-i did not support the use of violence. In one of our interviews, Bo-i described lashing out against local activists whose cause she supported, but whose methods she did not:

> You're turning into the people you're criticizing and worse. Mas marahas pa ang pamamaraan ninyo kaysa sa gobyerno. May penalty lang kapag 'di nakabayad ng buwis, sa inyo pag di nagbayad ang mga kompanya, six feet below the ground ang sagot. Mukha yatang mas masahol pa kayo kaysa sa (sistemang) gusto ninyong baguhin. When you say alternative dapat better than today.
>
> You're turning into the people you're criticizing and worse, you are becoming more violent than the government. When citizens don't pay taxes, they'll just get penalized. When companies refuse to pay revolutionary taxes, they end up six feet below the ground. It seems like your policies are worse than the system you're trying to change. When you say alternative, shouldn't it be something better than what we have today?

In 1989, a few years before returning to Mt. Apo, Bo-i began working with Attorney Solema Jubilan, a human rights lawyer who grew up in Kidapawan. In this capacity, Bo-i assisted Attorney Jubilan with a variety of projects, including documentation for Indigenous ancestral land claims, ensuring the delivery of social services, and advocacy for the respect of Indigenous human rights despite the ongoing conflict in Mindanao. Attorney Jubilan served as the legal counsel of the activists protesting against the geothermal project by the Philippine National Oil Company (PNOC) on the western flank of Mt. Apo.

In 1973, the Philippine National Oil Company was created by then-President Ferdinand Marcos to provide the country with a stable supply of oil and petroleum products, through oil and petroleum explorations.[18] In 1987, PNOC was given the additional authority not only over its existing geothermal operations, but also any future claims.[19] The Department of Environment and Natural Resources (DENR) initially declared PNOC's project near Mt. Apo illegal, but in 1991, they granted PNOC with an Environmental Compliance Certificate (ECC). The government's approval of the project was, in a way, a response to the economic damages brought about by the Gulf War in the early 90s, contributing immensely to the oil crisis. The economic impact of the oil crisis was further exacerbated by natural disasters such as El Niño, typhoons, and a major earthquake in Luzon, which occurred during the same period as the Gulf War.[20] These challenges provided the new president, Corazon Aquino, with an excuse to legitimize the fast tracking of energy projects on Indigenous land. One of the projects prioritized by the Aquino administration was the geothermal project in Mt. Apo, inspiring protests by local Indigenous peoples.

The PNOC filed a lawsuit against the protesters accusing them of inciting rebellion and subversion. It was through Bo-i's professional engagement with Attorney Jubilan that she was first introduced to PNOC. Shortly after she started working with Attorney Jubilan, Bo-i started getting death threats. As Bo-i became more deeply involved as an activist, she said that her dreams returned to Mt. Apo.

During this time, Bo-i began to travel further and further away from Mt. Apo, including attending rallies in Manila, more than 500 miles away from her community. Bo-i says that she consistently dreamt of her great grandmother, Omme. In the dreams, Omme was always searching for something: "Umpak ko, ingkon kos umpak ko? (My traditional clothes, where are my traditional clothes?)" she would ask. When Bo-i was a child still living at the foot of Mt. Apo, her great grandmother had been one of the few members in the community who still wore their traditional outfit. Bo-i dreamt of her great grandmother whenever she went to Manila. Bo-i interpreted these dreams as her ancestor's way of reminding her that she was drifting too far from her Indigenous community. Bo-i shared this conclusion with her cousins. They agreed with Bo-i; the dreams meant that it was time for Bo-i to return home. In 1995, Bo-i travelled from Kidapawan to her community at the foot of Mt. Apo. She vowed to continue her activism in Mt. Apo, working for her people-and-land.

Bo-i's belief in non-violent forms of resistance is best illustrated by her conversations with the Leftist organization her sister engaged with. Her interactions with her sister's friends not only introduced her to the social inequalities and abuses in Mindanao, it also encouraged her to take a stand. Her reaction to the methods of other members of the movement illustrated how she was reclaiming futures otherwise for herself and her political commitments. These early beginnings forged her path as an Indigenous leader who would later take space by reclaiming their Indigenous culture by using Indigenous knowledge as provided by her ancestors through dreams.

Dreams as a "hidden" source of cultural regeneration

> But here and there our sorrows grow
> Sometimes even the holy and the wise
> Chose to be mute, deaf and blind
> And the powerful and the mighty,
> Oh how they love to grab our lands.
> (Excerpt from the poem "Who We Dare" by Bo-i)

Between 1995 and 2016, Bo-i applied her experience as an urban activist to the struggle of her people-and-land, guided once again by her dreams, using Obo-Manobo beliefs and practices to strengthen bonds within the community and their resistance to outside aggression. Although Bo-i had grown up in the Obo-Manobo community, the 25 years she spent outside of the community had made her effectively an outsider. It took Bo-i seven years to earn her community's trust. At the end of this process however, she finally earned the title of *Bo-i*, reflecting her status as a respected female leader in the community. Since then, she has been known in the community simply as Bo-i, as I have referred to her throughout this chapter.

Soon after Bo-i's return to the community in 1995, she became involved in an effort to protect the Obo-Manobo's ancestral land from encroachment by the same utility company Attorney Jubilan fought against, the PNOC. The project proposed by PNOC was to build a 120-megawatt geothermal power plant which would displace Indigenous settlements in Mt. Apo. Their goal was to provide an additional source of electricity in Mindanao. In order to obtain the approval of the clans that would be displaced, PNOC approached Indigenous communities with a proposal for financial compensation.

Bo-i and her cousins saw that PNOC's proposals to members of various Indigenous communities in Mt. Apo had the potential to divide the Indigenous communities and remove them from their land. They agreed that they had to resist by unifying the many different communities within their clan, given that it would be difficult to unify the entire set of clans that lived around Mt. Apo. To unify the clan, Bo-i, her cousin, and the elders of their community identified two key factors: culture and bloodline. "Whether you're pro- or anti- the project, you'll dance your Indigenous dance, play the gongs and listen to *Uhwahing* (oral tradition)," Bo-i later said. A key part of this movement, she said, was the reincorporation of dreams as a part of Obo-Manobo community life. She named this initiative the Cultural Regeneration Movement (CRM).

The CRM emphasizes non-violent and non-confrontational forms of resistance, like building unity through cultural events. "Cultural regeneration is not a battle of times," she told a fellow Mindanowan activist:

> It is not a past brought to the present times nor is it a reinvention of past lives. For us, it is a process of reawakening of our traits, customs, traditions, and understanding of life that is buried deep within the hearts and memory of our tribe.[21]

In order to bring this initiative to life, Bo-i and her cousin, Tano Umpan, began planning a clan reunion and cultural celebration. Tano, in a conversation with Alejo during his fieldwork in 1999, said that he and Bo-i started the movement because of a dream. "It was there inside, deep inside, just waiting for the right time. When other companions arrived, then I thought it was the right time." Part of the regeneration Bo-i and Tano envisioned was to reclaim ancient ways of curing the sick through herbs and barks communicated to them through dreams by the spirits of Mt. Apo. Local healers from Sayaban and Pandanon joined the movement to help Bo-i and Tano pass on Indigenous knowledge of the healing powers of certain herbs from the forests of Mt. Apo combined with massages. According to them, Mt. Apo spirits visited them in their dreams to guide them on how to heal.[22]

They also used this movement as an excuse to gather members of the clan who they have not seen for a long time. "In our clan" Bo-i says, "our elders give importance to getting to know one's kin. There is a deep connection and attachment to our relatives." The first clan reunion spanned days and each day consisted of collaborative creation of genealogies and clan trees. Each family had to introduce themselves in the clan, perform and share a piece of their "forgotten" culture and essentially reconnect with one another. The connection was especially important to Bo-i and her cousin. One of the most important aspects of the CRM was the revival of communal conversations and dream interpretations. Bo-i does not remember specific dreams that were presented during the first clan reunion, but she does remember interpreting the dreams with the *boyahhon*. Bo-i described how previously-agreed-upon decisions were overturned because of the influence of dreams and the communal interpretation of those dreams. This confused and frustrated project officers from NGOs who were also trying to help the local Indigenous people, but Bo-i and her fellow leaders were confident that the alienation of outsiders was worth the risk. Indeed, the reunion was a success. The second reunion included nearly one thousand participants, including clans from all around Mt. Apo.

As a result of these and other advocacies, the CRM was perceived by the Philippine government as a threat. During the second clan reunion, participants noticed military men watching the festivities. Bo-i described the scene ironically: "They were waiting for us to proclaim, 'down with the state!'" Instead, Bo-i said that the military would have heard testimonies and stories. It took the many attendees several nights to introduce everyone and to share their personal knowledge of their culture. Every reunion was documented in detail by Bo-i.

Although the clan's resistance to PNOC was the primary catalyst for the CRM, the movement led to a broad resurgence in the clan's cultivation and education of its leaders. For example, Bo-i helped cultivate the next generation of Obo-Manobo leaders by sponsoring the education of selected Obo-Manobo students, who she considers her protégés. Bo-i refers to this group as her "second-liners." As she reported, she sees it as vitally important to her people's survival that they combine Indigenous knowledge with the formal education that can be gained from living and working in the city. Only by understanding how the city people work, she says, can Indigenous people use their knowledge to defend against future aggressions from outside their community to reclaim their land and their future(s).

In July 2004, Bo-i and her cousins finally obtained their land's Certificate of Ancestral Domain Claim from the national government. The eventual completion of the PNOC project at the foot of Mt. Apo has led me to believe that the granting of any certificate or title over ancestral lands by the State does not guarantee Indigenous autonomy.[23] Bo-i's resistance narrative exemplifies how Lumad people are marginalized through an aggressive national development agenda. Governmental use of nationalist discourses of "development" implies a forceful pressure on Indigenous communities like Bo-i's, to acculturate and assimilate to the rest of the Philippines compromising the futures of Manobo identity and people-and-land.

Bo-i continuously led her Indigenous community's resistance against development aggression and land exploitation by corporations and private actors. Bo-i experienced escalating threats from members of the private army whose superiors she previously confronted to protect her Obo-Manobo land. During this time, she was often visited by her dead brother through her dreams. One day, she dreamt of her brother inside a yellow taxi. Her brother asked, "bakit andito ka pa? Delikado buhay mo dito. Kung hindi ka sasakay dito, pumasok ka doon sa gate. (Why are you still here? Your life is in peril. If you're not going to ride with me, go inside the gate.)." According to Bo-i, the gate her brother pointed to was the gate of Father Alejo's workplace. When she approached the gate, Father Alejo's staff handed her a piece of paper. The next day, she went to Davao as instructed by her brother. Upon her arrival, Father Alejo's staff told her to look at the job opening on the bulletin board. The Philippine Canadian Environmental and Economic Management was looking for a community organizer and she seemed perfect for the job. Without a plan in mind, Bo-i applied for the position and was contacted right away.

Over the next few months, the threats against her that had been verbal became physical. Military men oftentimes searched for Bo-i in places she frequented, but she was always spared from the encounters. One day, while she was in Manila, she received information from her relatives and friends that in five different places in Mindanao, armed men came looking for her. In 2016, Bo-i was forced to flee Mt. Apo. She first sought refuge in Manila, and then in Canada.

As of 2019, Bo-i continues to reclaim Obo-Manobo futures using Indigenous knowledge and values as people-and-land. Bo-i's words and her life story are a testament of her active pursuit of knowledge, meaning, and desire to take her place in her Indigenous community despite being away for many years. Her ability to navigate through various realms and what role her dreams served in these social and political circles for her survivance as well as her people-and-land is admirable. Her being a Bo-i itself is proof that her ancestors and the spirits of Mt. Apo have accepted her as a leader and have allowed her to take her place in their Indigenous land. The spiritual and physical journey from the city to Mt. Apo and her eventual exile to Canada illustrate how Bo-i is being guided.

Dream life in exile

Today, dreams play an unexpected and paradoxical role in Bo-i's life, as the process of dream interpretation continues to evolve, as do the Obo-Manobo leaders who sustain it. Today, Bo-i is in political exile in Canada, more than 8000 miles away from her community. Because she is isolated from her people-and-land, she has relied on her dreams to help her make sense of her situation and her new role as a leader in exile. She relies on her dreams now more than ever. Ironically, Bo-i dreams more frequently about her home than she did while living in the Philippines. It is only by dreaming that Bo-i is present with her people-and-land by bridging her past reality to her present reality, and by feeling while remembering.

She is fortunate that she has managed to retain the community of fellow dream-literate people. It is by conversing with her relatives over the phone to share her dreams that she is energized. Bo-i also frequently interprets her dreams through poetry. One of her poems, entitled SILENCE, speaks to the feeling of separation she feels from her community. "Here lies a silence," she writes, "that crept into my flesh." To break that silence, she writes, her spirit separates from her body and returns to the land where she was born:

> Silence in a consciousness
> unwritten, almost unspoken
> in this part of the West
> thus here my body hides
> but my spirit soars or strives
> My body and spirit breaks
> into vast distance and time
> as I watch the winter skies
> as My Spirit returned to villages
> hearing Indigenous cries
> Everyday my heart departs
> Embrace you all in my past
> though engulfed with silence

to the fight I pay with full reverence
for a just peace in our land

Bo-i interacts frequently with local organizers in the Obo-Manobo community, but she has found that her dreams are not effective or meaningful ways to communicate with them. Bo-i described how the role of dreams has changed for the younger generations. Despite the broad success of the CRM, Bo-i said that the ultimate construction of the PNOC project changed not just the socioeconomic landscape of Mt. Apo, but also their clan's spiritual landscape. Even though Bo-i's clan found strength in the movement, morning traditions and rituals have been slowly replaced by TVs and gadgets, and the younger generation of Obo-Manobos are constantly pressed for time. As Obo-Manobo farmers left the fields to become labourers in the city, Bo-i reported that the farmers gave dreams less importance.

My conversations with Bo-i have led me to believe that this transition may only be temporary, similar to Bo-i's own time in Kidapawan City when she was forced to turn her back on her Indigenous roots. For 25 years, she did not engage in the dream interpretation and sharing that had so impacted her childhood, but the dreams did not abandon her. Instead, in time they led her back to her Obo-Manobo people-and-land, and they have stayed with her even in exile. It may be that her fellow Obo-Manobos are going through a similar journey now.

As of early 2020, Bo-i's role as an Obo-Manobo leader continues. She has relied on dreams for warnings, which she has relayed to her second-liners. Without her community's support and full attention, these warnings will just fade away. Bo-i has used poems, written in their Indigenous language and translated into English, to transform her thoughts and dreams into tangible forms. Her dreams have taken shape into words and these words' role continue to guide those who are willing to listen.

Conclusion: Dreaming of the future(s)

In Bo-i's experience, dreams have played an important role in the daily life of the Obo-Manobo. In the tradition of Vizenor's concept of *survivance*, I have argued that dreams give continuance to the stories of Manobo ancestors' people-and-land which play a vital role in Obo-Manobo resistance and the making of futures for the Obo-Manobo. In Obo-Manobo culture, dreams are meant to be shared and interpreted by the *boyahhon*. I have described how this process had a profound impact on Bo-i as a child, how dreams encouraged Bo-i to leave the city and return to her Obo-Manobo people-and-land in Mt. Apo, and how with Bo-i's help dreams were used as a vital part of their resistance against outside aggression.

Today, the Obo-Manobo are in an increasingly perilous position: Indigenous culture has been proven as an effective means of unifying their people in order to protect their people-and-land, but this knowledge is increasingly undermined by the encroachment of outsider knowledge. Dreams are a vivid illustration of this trend. Despite the challenges, Bo-i remains undaunted. She intends to continue working on behalf of her people. "Ang struggle na ito ay kaparehas ng sa aking mga ninuno, (This struggle is the same as my ancestor's struggle)" she said:

> Minana ko ito mula sa aking mga ninuno. Habang may taong nagpapasa, hindi kami mauubos. Sa ibang panahon, may ibang klaseng pagdurusa, but as a multiplier of good seeds, hihintayin kong tumubo ang mga binhi. In turn, mamumunga ang mga ito na may good seeds for the future.
>
> I inherited this from my ancestors. While there are still people who are willing to accept the challenge, we will not be extinct. In the near future, there'll be another form of suffering, but as a multiplier of good seeds, I will wait for the good seeds to grow. In turn, these will also bear fruits containing good seeds for the future.

Bo-i's dreams now live on paper – in her poems, and in this chapter. I hope that the reader will join Bo-i in the interpretation.

Notes

1 Lumad is a collective term used to refer to the Indigenous groups in Mindanao who resisted Islamization and Christianization. One of the Lumad groups which this chapter focuses on is called Manobo. Bo-i's specific community is the Obo-speaking Manobo of Mt. Apo.

2 Vizenor, Gerald R. *Manifest Manners*. Hanover: Wesleyan University Press, 1994.

3 Paredes, O. "Rivers of memory and oceans of difference in the Lumad world of Mindanao. *TRaNS: Trans-Regional and -National Studies of Southeast Asia, 4* (2), 329–349, 2016. doi:10.1017/trn.2015.28

4 Sentro Pangkultura ng Pilipinas, Manobo. Cultural Center of the Philippines Encyclopedia of Arts. Vol. II. Manila: Cultural Center of the Philippines, 1994.

5 Ibid.

6 Felix, Ma. Leny E., *Exploring the Indigenous Local Governance of Manobo Tribes in Mindanao*. Philippine Journal of Public Administration Vol. XLVIII, (1 & 2) (January-April 2004).

7 Rodil, B.R, and Alternate Forum for Research in Mindanao. *The Minoritization of the Indigenous Communities of Mindanao And the Sulu Archipelago*. Davao City, Phil.: Alternate Forum for Research in Mindanao, Inc., 1994.

8 Gerald Robert Vizenor, *Manifest Manners* (Hanover: Wesleyan University Press, 1994), vii.

9 Ibid.

10 Masinaring, Manggob R.N. *Understanding the Lumad: A Closer Look at a Misunderstood Culture*. Tebtebba Foundation, Philippines, 2014.

11 Paterno, M.P., & Oshima, N.M. Dreamweavers. Makati, Philippines: Bookmark, 2000.

12 Gaspar, Karl M. - *Panagkutay: Anthropology & Theology Interfacing in Mindanao Uplands (The Lumad Homeland)*. Quezon City: Institute of Spirituality in Asia (ISA), c.2017.

13 Miclat-Cacayan, Agnes N. *Babaylan: She Dances in Wholeness* http://www.babaylan.com/events/babaylan_symp_keynote_ANMC.pdf (accessed April 21, 2020)

14 Ibid.
15 Ibid.
16 Center for Babaylan Studies. https://www.centerforbabaylanstudies.org/history (accessed December 28, 2019).
17 Kidapawan was ratified as a City in 1998 by Republic Act No. 8500. Previously, it had been the provincial seat of Cotabato since 1973, and a municipality since 1947.
18 Lamberte, M. and Yap, J., *The Impact of The Gulf Crisis on The Philippine Economy*. Manila: Philippine Institute for Development Studies, 1991.
19 Ibid.
20 Albert E. Alejo, *Generating Energies in Mount Apo*. Quezon City: Ateneo de Manila University,, 295, 2000.
21 Ibid.
22 Alejo, Albert E. *Generating Energies in Mount Apo*. Quezon City: Ateneo de Manila University, 2000.
23 Gaspar, Karl M. - *Panagkutay: Anthropology & Theology Interfacing in Mindanao Uplands (The Lumad Homeland)*. Quezon City: Institute of Spirituality in Asia (ISA), c.2017.

References

Alejo, Albert E. *Generating Energies in Mount Apo*. Quezon City: Ateneo de Manila University, 2000.

Center for Babaylan Studies. https://www.centerforbabaylanstudies.org/history (accessed December 28, 2019).

Clariza, M. "Sacred texts and symbols: An Indigenous Filipino perspective on reading." *The International Journal of Information, Diversity, & Inclusion (IJIDI)*. 3, 2019. doi:10.33137/ijidi.v3i2.32593.

Felix, Ma. Leny, E., Exploring the Indigenous Local Governance of Manobo Tribes in Mindanao. *Philippine Journal of Public Administration* Vol. XLVIII (1 & 2), 2004.

Gaspar, Karl M. – Panagkutay: Anthropology & Theology Interfacing in Mindanao Uplands (The Lumad Homeland) – Quezon City: Institute of Spirituality in Asia (ISA), c.2017.

Lamberte, M. and Yap, J. *The Impact of The Gulf Crisis on The Philippine Economy*. Manila: Philippine Institute for Development Studies, 1991.

Masinaring, Manggob R.N. *Understanding the Lumad: A Closer Look at a Misunderstood Culture*. Philippines:Tebtebba Foundation, 2014.

Paredes, O. "Rivers of Memory and Oceans of Difference in the Lumad World of Mindanao." *TRaNS: Trans-Regional and -National Studies of Southeast Asia*, 4 (2), 329–349, 2016. doi:10.1017/trn.2015.28

Rodil, B.R, and Alternate Forum for Research in Mindanao. *The Minoritization of the Indigenous Communities of Mindanao And the Sulu Archipelago*. Davao City, PA: Alternate Forum for Research in Mindanao, Inc., 1994.

Sentro Pangkultura ng Pilipinas, Manobo. *Cultural Center of the Philippines Encyclopedia of Arts. Vol.* II. Manila: Cultural Center of the Philippines, 1994.

Vizenor, Gerald R. *Manifest Manners*. Hanover: Wesleyan.

6 Qishpikayqa aham

The hardships of becoming

Ñusta Carraza Ko

> It is an experience that those who have not lived through it will not understand…how the pursuit of education results in solitary suffering of *abandonar* (abandoning) the family and home.
>
> Francisco Carranza Romero

On many winter mornings of his birthday, my father would ask us to remember our own dates of birth. The dates were special to us, because father would speak in what seemed like a secret code at that time. "You must remember *Huk, Huk, Mana, Kimsa (One, One, Zero, Three)*—your birthday, and your sister's—*Mana, Kimsa, Kimsa, Huk (Zero, Three, Three, One)*." Smiling, we would repeat these numbers in Quechua and eventually remember them by heart. That is how I learned my birthdate, even before I knew how to write and say it in Spanish, the "official" language of our country, Peru.

These were not foreign words or a secret code. Growing up, we learned that they were Quechua, my father's mother tongue, one that he yearned we would remember and speak. He had held on to it, as part of his *lucha* (battle) for his *ayllu* (community that identifies as family per geography, bloodline, or shared objectives)—the Indigenous community of Quitaracsa in the northern highlands of Peru. The remembrance and practice of Quechua were his form of defiance against the dominant language of those in power in society, the language that he learned to have access to education and survive in society. Somehow through it all, his interests, work, and scholarship resulted in the nexus of linguistics and Quechua-scholarship, which helped him voice the stories of his *ayllu*. His experience was unfortunately, an anomaly.

More than seven decades later, the *lucha* remains. Despite the passage of legislation, such as the 2003 General Law of Education (Law No. 28044) that promoted "Intercultural Bilingual Education" and guaranteed the rights of peoples to access education in their mother tongue, the burden rests on the Indigenous population to maintain their languages while also learning Spanish—the language of the dominant majority society—as a second language.[1] The learning pertains to the Indigenous, an idea echoed in states such as Guatemala with majority Indigenous populations and bilingual educational policies, where the "development of Indigenous peoples" and the related work is "indebted to them, their own languages and cultures."[2] The urgency or will of bilingual education is absent for the majority population who speak Spanish and is not addressed in law. Indigenous peoples remain tasked with making the accommodations to claim space in society—one that through learning pushes for the loss of Indigenous languages, customs, and ultimately identity. These are symptomatic of the social, cultural, and political context of Indigenous experiences in society, that continues to ground itself in colonialist visions and places the Indigenous peoples not on a par with the majority population in power. Education reflects this view, setting the frame of the type of Peruvian society that is produced from this experience.

My father's stories of rural-to-city transition, the helpless feeling of forced integration into a classroom taught in a foreign language, and the realization of the absence of material goods within the Indigenous community are the stories of many others like him. It is a reflection of an Indigenous perspective about learning as a process of survival to sustain an Indigenous person and their *ayllu*'s life, and of survivance that involve "renunciations of dominance," loss, and "sentiments of tragedy."[3] This chapter discusses my father's journey for survival based on the temporality of his experiences of learning and loss. The hardships of displacement, recurring nature of the struggle, and repetitiveness of the discriminatory practices within education transcend linear conceptions of time. And thus, the memories are woven in a cyclical spiral time frame, where the stories of my father's past shape part of the present, influence the journeys of the past, and frame the futures to come. Understood within the context of learning, in my father's past, we see glimpses of the present and future stories of his *ayllu* and other Andean communities. The spiraling time did not disrupt the previous, current, and future marginalization of Quechua and other Indigenous language instructions, the socioeconomic status of Indigenous peoples, which remains unchanged, and the struggle for education continues for the Indigenous peoples of Peru.

The inevitable journey

"Javicho, *kashkita mikuy* (Javicho,[4] eat potato soup)." "Javicho, *anis yakuta upyay* (Javicho, drink anise tea)." "Javicho, *warayna* (Javicho, have a good night (sleep well))." My grandmother *Mama Pitushca* would often be heard saying these things to my father and his siblings, offering food, tea, and wishing them good night. At times, when someone in the community was in need, she would also make food for them and share what there was in the house to eat. That was the way things worked in Quitaracsa, where everyone shared, was a family, and supported one another.[5]

My father did not choose to leave Quitaracsa. Which child at the age of nine would want to do so? He wanted to be by his mother's side, like any other child of his age. He wanted to eat warm *kashki,* drink *anis* tea, and hear *Mama Pitushca*'s *warayna*. But he did not have a choice. If he wanted to help defend his *ayllu*, he had to become "educated." The education system hegemonized colonial logic in learning, wiping away differences in perspectives and identities, which included Indigenous knowledge and culture.[6] It was a microcosm of a society structured by colonialist visions of social class dynamics, with the state openly excluding and penalizing Indigenous people for "behavior which deviates from those adopted by the dominant legal and cultural system," referring to Indigenous customs, traditions, and ways of life.[7] And yet, my father understood that the way to fight the system was to understand the ways of society and object to the injustices with the language and ways of those in power. Society provided no other option of access point to society, so he forcedly took the chance to seek "education."

At a young age my father had seen what had happened to his community and the community leader—my grandfather *Taita Matildo*. Quitaracsa was a communal land. As *Taita Matildo* had often said, it was the "land of those who worked for it." However, such communal understanding had been disputed by strangers, mainly city-people, who claimed landownership. One could argue that this territorial aggression was an Andean-extension of the 1893 Law of Immigration and Colonization and the 1909 General Law of Mountain Lands that permitted the land grabbing of the Amazonian Indigenous peoples' lands by settlers and those involved in the rubber industry.[8] On the other hand, the territorial disputes in Quitaracsa were taking place after the passage of the 1920 Peruvian Constitution, which dictated under Article 41 that the collective rights to land belonging to Indigenous communities were imprescriptible and could only be transferred via government securities.[9] But that law that protected Indigenous peoples' rights could not compete against the colonialist rationale that dominated the Peruvian society. Hence, land grabbing persisted throughout the 1920s through the 1940s in Quitaracsa. Those who threatened the livelihood of Quitaracsa were not related to any *Quitaracsinos* (people from Quitaracsa), spoke no Quechua, and were not from neighbouring Indigenous communities. And yet, they had decided that the Indigenous lands were there for them to take and registered the territory under their names. Territorial possession was a foreign concept to the community, as land was a sacred matter, and belonged to everyone. This was a shared concept among Indigenous peoples from other regions such as Guatemala, where communal work was the centrepiece for the production and protection of the community's common interests.[10] These ideas had therefore influenced the reason why Quitaracsa's land had never been registered. That is, even after the Peruvian government had begun processes of titling communal lands in 1928. The news from Lima took a long time to reach rural communities. Given the few numbers of Spanish-speakers who could read about the titling and would know how to go about soliciting the paperwork, land registration was never considered in Quitaracsa. The cultural difference in the conception of land ownership and the lack of resources for Indigenous communities to quickly register their lands was the loophole through which Quitaracsa and other neighbouring communities lost their communal ownerships.

The documents proving foreign ownership were generated in public notary offices from the city of Caraz, the capital of the province of Huaylas in the Ancash region. They were all written in Spanish, a language that not everyone knew how to read or write in Quitaracsa. At an early age, my father witnessed how outsiders would come through Quitaracsa and claim property ownership with papers written in Spanish. No one had consulted *Quitaracsinos* of their rights to land and now the land was in the hands of strangers who knew how to navigate the system and speak Spanish. Such were decisive moments that shaped my father's life. It mattered to him, *Taita Matildo*, and the community that there be someone who would learn the ways of majority society and know Spanish well enough to help defend the interests of Quitaracsa.

Other related critical junctures came during the early years of my father's birth. In 1945, *Taita Matildo* had gone to Lima to petition for the creation of a school in Quitaracsa to the Ministry of Education. After a year-long wait a resolution was issued by the Ministry that authorized the establishment of a primary school. However, the authorization was for the creation of a school offering schooling up to the second grade. At the time, the 1941 Organic Law of Public Education (Law No. 9359), passed during the first presidency of Manuel Prado (1939–1945), was in effect.[11] According to the law, the state was responsible for distributing education without distinction, to all Peruvian people. However, the same law included provisions that would limit the type of education that rural and Indigenous communities would receive. For instance, Article 122 noted that primary schools established in rural areas with predominant Indigenous-language-use would need the most basic study curriculum and educational programs. The purpose of these schools was not comprehensive education. The focus of these schools was the "teaching and learning of Spanish," or the *castellanización* of the people.[12] Accompanying Spanish instruction, students would be subject to the "adoption of civilized customs," to integrate them into the majority culture.[13] What was civilized was not Indigenous, an idea that contradicted the first Constitution of Peru in 1822, where at least in text all Peruvians were considered equal before the law.[14] But even at that time, in discordance with the Constitution, Indigenous peoples were not given equal rights. For instance, electoral laws in Peru (until 1890) did not authorize voting for the illiterate population, the majority of whom were of Indigenous descent.[15] And, Indigenous peoples as a group were legally recognized for their collective rights only during the governments of Nicolás de Piérola (1895–1899) and Augusto Leguía (1919–1930).[16] The schools and the Organic Law on Public Education reflected the historic racist views against Indigenous peoples by focussing on propagating the ways of the majority culture, customs, and language, in efforts to "civilize" the population. In the process of incorporating the Indigenous person in society, they needed to be "educated" and "civilized."[17] Of course, this was with the caveat that both language education—largely an alphabetization project—and "civilizing" processes be wrapped up in two years of school. Reflecting these arbitrary criteria of elemental education, a provisional school was created in Quitaracsa in 1947.

The school was basic with an emphasis on Spanish education. Still, it was a noticeable change for the community, a welcomed move for *Quitaracsinos* who previously lacked access to any educational resources. But, *Taita Matildo*'s efforts to petition for the establishment of a school in Quitaracsa had not gone unnoticed by those in power who feared change. By order of one of the many strangers who claimed to be the *hacendado* (landowner) of Quitaracsa, *Taita Matildo* was detained, accused of fomenting a rebellion among the people, beaten, and locked up in a cell. The *patron* (head of the hacienda) had heard of *Taita Matildo*'s efforts to create a school. And, he was determined that education was not needed for Indigenous peoples. He claimed that these were Matildo's efforts to make Indigenous peoples conscious and make them no longer obedient to the *patron*. Education was going to challenge his position in power as reading and writing would make Indigenous peoples understand their rights, which included the equal rights guaranteed to them as Peruvian citizens under Article 4 of the 1933 Constitution of Peru.[18] The possibility of class conscientization was enough to give the *patron* a reason to mistreat Matildo. And, so it was arranged that Matildo would be detained, brutalized, and be sent to a prison in Caraz. What the *patron* did not know was that Matildo had already gotten approval for a school from the Ministry of Education. And, to the dismay of the *patron*, a nightguard at the *hacienda* had heard of the story, pitied Matildo's battles for education, and cracked open the cell door in the middle of the night to allow my grandfather to escape.

The treatment from *hacendados* and local authorities, who my father remembers as the strangers who came to Quitaracsa speaking a different language, reflected a negation of Indigenous peoples' access to opportunities in society which was reserved for those who were part of the majority—the Spanish-speaking population who were of *mestizo* (persons of Indigenous and Spanish mixed origin) or *criollo* (persons of Spanish or European origin) backgrounds. *Taita Matildo* was fighting against a system that was engrained to provide a structural difference based on ethnic and socio-economic class relations from the colonial period. Intrinsic in the relationship between Indigenous and non-Indigenous populations was the association of "class stratification with ethnic identity, particularly of whiteness with higher economic classes," and the vision of Indigenous peoples as being retrograde and "uncivilized."[19] In contrast, the non-Indigenous population was associated with "progress," development, and the standard of what constituted to being "civilized."[20] Such views were commonly shared in contexts of post-colonial states, where Indigenous peoples were regarded as a problematic population who prevented states' modernization processes.[21] In the struggles of *Taita Matildo* my father noticed these dynamics of class, ethnicity, and language. And, it further made the little boy motivated to pursue his studies.

Although he was just one person, my father understood that pursuing education and making an entry into society as an Indigenous person could help his community navigate the social and political system and ultimately protect communal interests. It was his way of halting the colonial project of dispossession directed towards the Indigenous peoples. As *Taita Matildo* had told my father, education would be a way of fighting for his *ayllu*. But the journey to the city of Caraz, the closest metropole of the region, was a long road away from *Mama Pitushca* and Quitaracsa. Still, it was an inevitable journey, the first of many that my father would have to make to pursue education. At the age of nine, he set on foot to Caraz, away from his mother's warmth and the comfort of his home.

La escuela *(The school)*

When he arrived to Caraz, my father had finished the second year of schooling in Quitaracsa's provisional elementary school. Fulseda Caballero Rincón had been the schoolteacher in Quitaracsa. She, along with Pancho, my father's older brother,[22] had convinced *Mama Pitushca* and *Taita Pitushca* that Javicho had to study in a big city like Caraz. She had seen his potential to continue education. There were schools in neighbouring Sicsibamba, Pomabamba, and Yuramarca districts, but they lacked resources and opportunities that Caraz had as the capital of the Huaylas province. Additionally, Caraz would make sense as the next place in my father's journey as *Tio Pancho* and the teacher Fulseda, who became *Tia Fulseda*, were going to settle in the city. So, while he felt alone in a new city-setting, the rural-to-city family transition had not been too difficult. What was difficult was to feel the transition being emphasized in the educational setting, with teachers who would treat him and others like him differently, by sidelining them for their appearance and manner of speech.

The teachers were aware that my father had come from outside Caraz. They knew that his mother tongue was Quechua and that he had learned Spanish in a provisional primary school. At the time, Spanish was the single official language of Peru. Although Quechua was the most spoken Indigenous language, it was only recognized as the other official language of Peru via Decree Law No. 21156 in 1975, during the administration of Juan Velasco Alvarado (1968–1975).[23] Given this background, hence, the teachers were not trained to guide the learning and accommodate the linguistic differences of Quechua or bilingual students in the classrooms. In fact, as Caraz was not an Indigenous community there was no need as laid out by Article 42 and Article 128 of the 1941 Organic Law of Public Education for teachers to have knowledge of Indigenous languages to instruct students.[24] Nonetheless, the absence of Indigenous language emphasis by teachers in Spanish-speaking areas was not a blanket excuse for teachers to mistreat students. But the behaviour was rampant, and the practice went unchallenged. The Organic Law did not deal with such discrimination matters. In fact, if the behaviour was within the context of disciplining students, under Article 100,[25] teachers were free to discipline students with the consideration that the pupils' liberties of expression were respected. Simply put, the system was positioned to allow discrimination and students like my father were singled out.

It was as though the memory had left a scar. My father remembers every detail of what happened that day, when he turned in the exam. Receiving the exam, the teacher laughed and pointed out the question he had not answered. The discussion revolved around a word he did not know in Spanish. The exam question asked for students to answer which were the "*oriundo* (native)" plants that the Incas used to eat. The word "*oriundo*" is not a commonly used word even for native Spanish-speakers. Nevertheless, the burden fell on my father to answer the exam question. When he turned in the exam without a response, the teacher shamed my father for not knowing the Spanish word. Then, the teacher ridiculed him for not being aware of what the Incas ate. It was in front of the whole class. The implicit reference from the teacher was that since my father was Indigenous, he ought to have known what the Incas used to eat. This classification of identity based on ethnolinguistic grounds derived from the periods of colonial government, during which as conquistadors and colonizers of Peru, the Spanish and those of Spanish descent ruled with the idea of ethnic and linguistic superiority. Those of Spanish descent who immigrated or were born in Peru "dominated the upper class, congregated in urban coastal areas, and looked to Lima and sometimes to Spain for their identity."[26] Their linguistic identity was tied to the Spanish language. Comparatively, the Indigenous peoples belonged to the lower classes of society, resided in Andean rural highlands, and "looked to Cuzco, the ancient heart of the Incan Empire, to represent their national identity."[27] Their linguistic identity was defined by Quechua. The classification of ethnic and linguistic identity enforced a hierarchy, where anything Spanish-related was superior and the Indigenous was not. This was evident in the teacher's attitude. My father could not have known all the Spanish words. He was a third grader who had learned Spanish as a second language. Additionally, being Indigenous did not guarantee my father's ability to know the culinary customs of the Incas. My father could have been from Puno, an Aymara-speaking part of Peru, and not had a clue about Incan traditions. Historically, the Aymara population were one of the various minority groups that were brought under Incan rule in the 15th century. Although they were under the reign of the Incan empire, the *Aymaras* continued practising their own linguistic and cultural traditions and resided in the southern Andean regions of Peru and Bolivia. What had taken place was a discrimination from the teacher towards my father, a Quechua-speaking student of Indigenous origin. She purposefully shared with the class my father's question omission, pointed out what she considered was a linguistic incapacity, forcedly associated his Indigenous identity with the history of the conquered, and had humiliated him. These were the types of interactions that would embed subtle and overt moments of discrimination from teachers.

The mistreatment clearly went beyond a matter of language education. Even if my father had proficiency in Spanish language writing, reading, and speaking, the attitudes would not have changed. The social perception was rooted in far more material matters, which had resonance to colonial perceptions of ethnicity and class. The Indigenous were the poor and the higher classes were occupied by those of Spanish descent or of mixed Spanish origin. Implicit in this idea was the association of whiteness with a betterment of society, one that was echoed in the nationalist project of many Latin American states under the disguise of *mestizaje*.[28] Such views manifested in the *indigenista* movement of the 1900s—a cultural movement influenced by political conversations about Indigenous exploitation and marginalization in society—with some scholars arguing that the "solution to the Indian problem," or "Peru's problems of economic and national integration was to improve the race through immigration."[29] The same logic influenced immigration policies that invited European but discouraged Chinese and Japanese migration to Peru in the 19th century and set the undertone for eugenics-based policies that coercively sterilized Indigenous peoples from 1996 to 2001.[30]

When my father moved to Caraz to study, his family supported him financially to the best of their abilities. The needs of a child studying in a city, however, were quite different than what *Mama Pitushca* and *Taita Matildo* expected. At the time, my grandparents were able to accommodate my father's stay in Caraz with basic necessary provisions, such as food, medicine, and books. However, they were not at a financial point to overspend, particularly on material goods such as fancy clothes or shoes. So, my father went to school without shoes. What *Mama Pitushca* and *Taita Matildo* did not know was that as a child coming from an Indigenous community, my father's attire and appearance would be more scrutinized by society. In Caraz, it mattered for you to wear a set of "proper shoes" and "dress in an appropriate manner," as the teachers had explained. The implicit reference was to wear clothing styled the European way. What constituted the "correct" appearance, provided a glimpse of the historic relationship between the colonial state and Indigenous population, fixated on physical appearances. Clothing was an external symbol of identity. Hence, to discard clothing with traditional Indigenous textiles and use Western attire, implied replacing a cultural marker "considered developmentally previous and inferior" with something better.[31] Clothing indexed discrimination. The Western attire became an "important component of the civilizing process,"[32] and what was Indigenous, whether it was clothed or not, became a symbol of inferior culture.

Not knowing the societal dynamics that defined Indigenous ethnicity, class, and material goods in a vulnerable position to that of the majority culture, my father began the schoolyear. He was one of the few students who did not wear shoes to class. He did have books, notebooks, pencils, and basic clothes to study. But he was missing the shoes, which at the time were pricier than other goods. The unintended "statement" as my father recalled that he made with his "no-shoes," made a difference in how he would be treated in class. He was excluded from outdoor school activities. This included the school's official marches around town—where all students participated. The teachers specified that they could not put him to march along with other children, as he wore no shoes and did not have the appropriate clothing. Such words resonated with the colonial perspective of associating the Indigenous clothing style with non-civility. As a young child, still trying to figure out how to follow the curriculum, adapt to a city-like environment, and make friends, these were difficult experiences. But even at that age, my father understood that some things were done "purposefully" to "*tumbar* (make fall)" those who were different. His identity of being from a rural Indigenous community and of being a Quechua-speaker were markers to push him aside from others. The absence of material richness only exacerbated the situation.

La escuela (the school) became a microcosm of Peru's societal dynamics. Those of Indigenous background who spoke Quechua were marginalized from others who spoke Spanish and claimed a stronger connection to their *mestizo*, Spanish, or European roots. This was a new form of "othering" that my father experienced. Although the interactions with the *hacendados* in the past had given him a taste of this treatment, in educational settings, whereby attending the school he had signalled his willingness to adapt to the majority culture, he was caught by surprise. But, this was an inevitable outcome. The educational system was built upon colonialist visions of differentiation, where "the Indigenous" was perceived as part of the "pre-history of the nation," as a group or an identity that needed to be "overcome," to reach progress.[33] Explained in the context of time, colonial rationale had transcended time, adapted to the new context, and morphed into institutionalized discrimination. My father faced these remnants of the colonial past, a version adapted to the 20th century.

The separation also included elements of economic class differences, with my father's experience exemplifying the treatment towards those who were Indigenous. The school imposed on individuals like my father to accept the treatment they were given, and assimilate and learn the language and customs of the majority society. Again, these were reflective of past colonial policies of class differences. My father did learn Spanish and accommodated to the way things worked in a Spanish-speaking city environment. However, the experiences at *la escuela* helped him define what mattered more to him. To defend others like him and his *ayllu*, he had to continue his *lucha* and hold on to his identity, language, and customs.

Temporality of learning and loss

Before his death, *Taita Matildo* gave each one of the family members a woolen sling shot he made with zigzag patterns. He explained to us the importance of the zigzag, particularly on how each pattern connected with the others and how each zigzag, despite its shared structural points with others, had new inputs. This was his gift, one that he believed would represent the coexistence of place, time, and space. Growing up with such influences, I learned that in the Andean world, time is not a linear concept. Time and space are in a circular spiral, in that the past connects us to the present and yet carries its legacies and continues to exist as part of the present. This does not mean that the past repeats itself exactly the same. Rather, each cycle of time that returns to the past always finds itself with new protagonists and contexts that allows for growth and change. The experiences are in cycles of journeys, moving forward, renewing, completing, and folding back. In this spiralling time, those of the past live among us in memory and spirit and become a part of our present journey. The temporality of time does not limit or separate our existence from the past.

My father's multiple travels back and forth between Caraz and Quitaracsa were cycles of journeys. They embedded continuous experiences of learning and loss. Although each time, the learning and loss varied in magnitude, context, place, and age. He first left to Caraz seeking education, leaving his home behind. When he thought he had progressed to a new level in his studies, he went back to Quitaracsa and reconnected with his beloved mother. The stay would be temporal, as *Taita Matildo* and *Mama Pitushca* reminded my father that the time had not come yet for him to stop his studies. He had to continue, to make the best of the opportunities presented to him and become the voice for his *ayllu*. Leaving *Mama Pitushca* once again, my father would recomplete his trip back to the city. The few trips would turn into multiple journeys, spanning years of studies and travel. The final destination cities would change, time would move on, and he would finish college, graduate school, and eventually become a professor. But in this spiral of returning seasons, some things remained the same. The scenery of houses made of adobe with *techo de paja* (straw roof) would greet his eye and the smell of *anis* would fill *Mama Pitushca*'s kitchen. Other things did not. *Mama Pitushca* grew ill and would unfortunately disappear too early from the scenery, a loss that my father could not prevent. That was the "blow" as he called it, that my father had to take. To complete his studies he traded in the moments of being with family. Time was circular, turning back and turning in, and the journeyed paths he had taken were part of his academic transformations and his work on becoming the voice for his *ayllu*. *Mama Pitushca*'s loss would too become a part of a memory that he would take in pursuing works related to his community and Quechua-linguistics further. And, eventually the little boy that set on foot to Caraz would remember the journeyed paths with love, or as he called it, "*hasta sentirle cariño como a un hermano* (until reaching the point of feeling a brotherly love)."

After his studies in primary school, he had joined the seminary, where he knew he could be provided housing and education, and perhaps in doing so, go back and help his *ayllu*.

The seminary was rigid. The curriculum was fixed, and it was difficult to travel back to Quitaracsa as often. The language of instruction was entirely in Spanish. And, there was barely any time left to practice Quechua or as previously noted, go back home. It was also in the seminary that my father learned Latin and Greek, which broadened his understanding of Spanish as a language. He could now read, write, and teach Spanish at a higher level than the teachers in primary schools. After all, now he had the basis of Latin, which was the root of romance languages.

The education paths supported both by public and religious institutions strengthened, as well as expanded the breadth of his Spanish.

Interestingly, it was this heavy emphasis on Latin and Greek, and insistence on Spanish-only curriculum that allowed him to draw on Quechua in comparative analysis. He had not become someone else by being in the Spanish system. There was no loss. Rather, he had gained the tools to be able to analyze and critically reflect on Quechua and Spanish even further. The seminary opened up the possibilities for more in-depth education in the arts and humanities, which he used to translate short stories from Spanish to Quechua and vice versa. These became part of the collection of stories about Andean mythologies later published in Quechua, Spanish, and translated into Korean.[34] Through the stories, he breathed air into his *ayllu*. Many of the *Quitaracsinos* participated in the process as storytellers of their past and present. And so, even after their passing, their memories continued to live in text, in the pages of publications, and in the minds of other Quechua, Spanish, and even Korean speakers. In my father's stories and works, his *ayllu* of the past continued to exist in the present and in the possible future.

The formative years in the seminary also helped my father learn the need to rebel against the hierarchy and defend his rights, both as an Indigenous person and as a person from a developing country. He, along with his colleagues, rebelled against the North American priests who were trying to impose their views on their educational curriculum, by for instance negating the seminarists' outreach to neighbouring Indigenous communities. The consequences of those actions resulted in his ousting from the seminary. Instead of being a setback moment, this became an opportunity for him to pursue university education in the study of languages with a linguistics major. And, with that education, he published one of the first Quechua-Spanish dictionaries. The book was his way of paying back to his *ayllu*, to be a guide for those who could not speak Spanish but could read Quechua. It was also a way for him to claim space in academia, dominated by *mestizo* and *criollo* scholars and their publications on Indigenous languages, identities, and cultures. My father was writing as a person from the Indigenous community, taking his place in the field of Quechua-Spanish linguistics and in doing so, providing a voice for many Indigenous *ayllus* that were forgotten. The personal losses he had suffered in learning had renewed themselves in new opportunities for engagement, and through the publications, he shaped other futures to come. I, the author, am a testament to his influence.

La lucha *(The fight)*

Many years have passed since my father left home to pursue his education. Reflecting the change in time, on the legislative front, the state has accommodated drastic revisions of its stance towards Indigenous communities and education. Peru ratified various international legal instruments that recognized the right to education and the rights of Indigenous peoples. Namely, Peru ratified the Convention Against Discrimination in Education on July 16, 1963, the International Covenant on Economic, Social, and Cultural Rights (ICESCR) in 1978, and the International Labor Organization's Convention

169 in 1993. Under Article 1 of the Convention Against Discrimination, "discrimination" constituted any "exclusion or preference" based on race, national or social origin, and economic condition or birth, among other things, that had the "purpose or effect of nullifying or impairing equality of treatment in education."[35] These included limitations imposed on any group of persons to "education of an inferior standard."[36] The same was true for ICESCR. Article 13 of the Covenant outlined the right for everyone to education. Under Section II of Article 13, the Covenant specified that primary education "shall be compulsory and available free to all," secondary education made "generally available and accessible to all," and higher education made "equally accessible to all."[37] On related educational grounds, Convention 169's Article 26 and 27 discussed the necessity for the Indigenous peoples to have the "opportunity to acquire education at all levels on at least an equal footing with the rest of the national community," and the necessity for the educational program to incorporate Indigenous peoples' histories, knowledge, and value systems.[38]

Looking at the 1941 Organic Law of Public Education retrospectively from these normative standards of the Conventions and Covenant, the provisions laid out in the Organic Law were highly controversial. They differentiated the access to equal education, by restricting Indigenous communities to primary education. Then, the Organic Law emphasized the most basic educational curriculum for Indigenous communities, a discrimination based on ethnic grounds in not providing education for Indigenous peoples at an equal footing with the rest of the population. These ideas were at odds with the international norms from the Conventions and ICESCR. The state's adoption of the international norms parted the state from the standards of the Organic Law that marginalized Indigenous peoples. They signalled a progress in the state's position to make education more equally available to the Indigenous communities, disregarding their ethnic or social position.

Reflecting the adoption of such international documents which upon ratification form a part of national law, legislation related to Indigenous peoples and the state's input on Indigenous education have also improved.[39] For instance, language policy changed from a monolingual to a multilingual state policy with a regime change from an oligarchic state rule during the 19th and early 20th century to a more "progressive-minded military regime in the 1970s" with Velasco Alvarado (1968–1975) whose reformist policies sought to "empower rural and urban lower classes."[40] As part of the education reforms, in 1975, Quechua was recognized as the official language of Peru. This move added value to the language of the majority Indigenous population. Such changes were accompanied by other reforms, such as the 1969 agrarian reforms that aimed to dismantle oligarchic rule based on *latifundio*—large estates owned by a single family or an individual that were only partly put to use—by implementing land redistribution and putting the production of land in the hands of the *campesinos*. The new policies, both in education and agrarian sectors, were unsustainable in a society built on a system of hierarchic differences of class, ethnicity, and power. In a few years, Quechua was downgraded with the 1979

Peruvian Constitution that specified the use of Quechua and Aymara, primarily in certain geographic zones. And, while the agrarian reforms expanded the *campesino* class, it also created a class identity which rather than overlapping with ethnicity, instead replaced the ethnic identity of individuals, with Indigenous peoples now referring to themselves as *campesinos* and not as Indigenous. Despite the reforms' shortfalls, Velasco Alvarado's government at least signalled the beginning of major changes in Peru. Most importantly in the linguistic front, Quechua was recognized as a national language of Peru, which was a symbolic and critical moment that helped elevate the respect for Indigenous languages.

Adding to the impetus of change, during the 1980s, the state initiated the official formation of teachers specifically for Andean region education. This was part of the movement that promoted both bilingual and intercultural education as "liberating education," moving away from assimilationist policies of using Indigenous languages to "transition to Spanish," and promoting the maintenance of Indigenous languages and the acquisition of Spanish as a second language.[41] This was part of a long history of bilingual education training that was initiated in 1945 with the partnership between the *Instituto Lingüístico de Verano* (ILV: Summer Institute of Linguistics) and the Ministry of Education, to train teachers to help educate the Amazonian area of Peru. The ILV was actually an American Christian organization, which at the time used its "*cobertura academica*" (academic front) affiliated with the University of Oklahoma that allowed the group to officialize its partnerships with governments.[42] Program objectives included the development of a literacy program for Indigenous communities in the Amazonian region of Peru with an emphasis on Spanish education. The other objectives were the promotion of Spanish and bilingual education with the aim to "acculturate" Indigenous peoples and translate the New Testament and other religious books to Indigenous languages that would facilitate evangelization.[43] Acculturation for the ILV—that worked with governments to include them in the nationhood—was understood as the assimilation of "inferior ethnic groups," referring to the Indigenous population, into the Spanish-majority-culture via language education.[44] ILV also used education for its mission work. The evangelization impulse was built on colonialist visions of hierarchy, with the ILV placing itself in a superior position from which to educate teachers. These teachers would then "help" the Indigenous peoples as they were deemed as needing knowledge and salvation. In the process, the customs and belief systems of the Indigenous were discarded. Such ill-fated intentions from the ILV were documented in other countries with large Indigenous communities. For instance, in Guatemala, ILV disseminated evangelization in Mayan communities, by connecting their work to the oppression of the Mayas, focussing on development works of education, healthcare, and agricultural programs.[45] The mission of the ILV was a different form of colonization, which in the name of fomenting "education," "literacy," and providing philanthropic work, also oppressed Indigenous peoples by negating the Indigenous inclusion in educational program creation. For this reason, the ILV was viewed with skepticism from Indigenous communities who regarded the institution as threatening their "traditional values."[46]

The colonialist evangelical intentions of the ILV enabled the imposition of foreign beliefs and a forced acculturation of Indigenous peoples. But, despite these problems, the cooperation between the ILV and Ministry of Education established, at a high cost, some form of acknowledgement of the value of promoting bilingual education initiatives and the importance of training teachers in Indigenous languages. In that, it is worth noting that some respect for Peru's pluricultural societal context was established. For instance, the cooperation provided the basis of the establishment of the *Instituto Superior Pedagógico Bilingüe* (Superior Pedagogical Bilingual Institute), which became the starting point for region-specific institutes dedicated for bilingual education. These included the Superior Pedagogical Institute of Puno and the Superior Technological and Pedagogical Institute La Salle in Urubamba that in 1989 created educational programs to train teachers bilingually to teach in Quechua or Aymara-speaking neighbourhoods in Puno and Cusco regions.[47] Along with these regional institutes, in 1989, the state established the General Directorate for Bilingual Education in the Ministry of Education.

Continuing the emphasis on bilingual education, from 2000 to 2002, the state also added an element of intercultural education training for teachers.[48] And in the last decade, the Ministry of Education created four "intercultural universities"—the National Intercultural University of Amazonía (UNIA), the National Intercultural University Fabiola Salazar Leguía (UNIBAGUA), the National Intercultural University in Central Jungle Juan Santos Atahualpa (UNISCJSA), and the National Intercultural University of Quillabamba. The universities foster an environment favouring "intercultural coexistence" and the "equality of access to educative opportunities" for Indigenous peoples.[49] They reflect the state's recognition of the pluricultural background of Peruvian society, and the acknowledgement of the rights of Indigenous peoples enshrined in international and domestic legal documents. The mission of these universities incorporate Indigenous peoples' conceptions of wellbeing and development, along with their cultural values and traditions. Of the four universities, UNIA is the only institution with registered students.

While the creation of the General Directorate, *Instituto Superior Pedagógico*, and even intercultural universities could be regarded as wins for Indigenous communities, there were many shortcomings that were reminiscent of the past. First, government initiatives dedicated for the promotion of bilingual education faced pushback. The General Directorate of Bilingual Education was forced to cease its operations in 1992 during Alberto Fujimori's presidency (1990–2000). When it re-opened in 1996, the office underwent a change that uprooted its original objectives. From a directorate focused on bilingual education, the office transitioned to a directorate serving rural education. The newly defined objectives of the General Directorate of Rural Education resulted in the sidelining of intercultural and bilingual education in both primary and higher education settings.[50] And, even with the changes in government that created the General Directorate of Intercultural, Bilingual, and Rural Education, there is a lack of connection between the policy positions of the Directorate and the implementation on the ground. In fact, within the Ministry of Education only those who work in the General Directorate and bilingual scholars who advise the office seem concerned about promoting intercultural bilingual education.[51] Hence, even with restructured government initiatives within the Ministry of Education on intercultural bilingual education, the challenges remained in making Indigenous language, culture, and education a priority for the state.

Second, objectives of intercultural bilingual education were not predicated on mutual learning. As noted previously, the emphasis was on the maintenance of Indigenous languages, the propagation of Spanish to the Indigenous population, the assimilation of the Indigenous peoples to the Spanish-majority-culture, and even evangelization. The policy was thus, unidirectional and colonialist. Intercultural education did not include the learning or promotion of Indigenous languages to the Spanish-speaking peoples. Rather than achieving equity, the goal was to sufficiently get the Indigenous population to communicate with the rest of society, not a dialogue or learning among different cultures. In this process, the Indigenous peoples would also become assimilated to the majority culture. This was also a focal point of criticism from Ecuador and Bolivia's Indigenous organizations against their respective states' intercultural bilingual education. The groups questioned their states' intercultural bilingual education as depreciating Indigenous cultures and serving the purpose of Spanish-language-transition. They likened the intercultural policy as missing the elements of respect, "mutual learning among cultures," and categorized the objectives as more of an anti-discriminatory policy of inclusion.[52] Such criticisms were also echoed from Indigenous community groups in Peru against the ILV's bilingual education program, which they likened to a new linguistic-education-based colonization program that disregarded Indigenous cultures and enforced evangelization.[53] And, what about the intercultural universities? The pursuit of higher education based on "intercultural" objectives had its limitations. For instance, at UNIA, Indigenous leaders and teachers were excluded from employment. Instead, the university brought in *mestizo* professors from the National University of Ucayali (UNU) who did not "have the intention to implement an intercultural" education.[54] The tensions erupted in 2010 with protests of UNIA's Indigenous student leaders who accused the administration and professors of racism and discrimination against Indigenous peoples. This was reminiscent of my father's stories as a young child, fighting against teachers' discrimination in Caraz.

Moreover, bilingual education has not been emphasized beyond the training of teachers. There has been no national-level curriculum change which includes bilingual education or language immersion programs. Government documents indicate that there is still a strong resistance to developing educational policies that recognize the value and importance of Indigenous language education and Indigenous values and customs.[55] Indigenous communities have long demanded that Peru model its system to that of Bolivia, which provides an education system of three languages—the mother tongue, one national language, and a foreign language course.[56] The state has not responded. In fact, universities that have implemented programs for Indigenous peoples (e.g., National University of San Marcos, National University of San Antonio Abad, or the National Agrarian University of Tingo María) have done so independently, without the state's support. Most of these initiatives have been to "reserve some space" for Indigenous students' undergraduate admission.[57] Students from Indigenous communities interested in attending these universities apply and compete with other Indigenous students for a spot in the entry, allowing for a fairer comparison separate from Spanish-speaking students who did not face linguistic difficulties or lacked resources to access education. However, once the students are admitted to the university, they are left on their own to navigate the system. Even when a student from an Indigenous community is given a government grant to pursue university studies, there are no follow-up guidelines from the state to help accommodate the student to the new environment.[58] In other words, while the initiative may be there to bring in Indigenous students to higher education, there are no other mechanisms to guarantee their retention. Nor are there national educational programs with strong emphasis on bilingual education.

And, finally, education, as had been in the past, remains difficult to access. The distance problem persists. Main educational centers remain in the cities, away from Indigenous communities.[59] Quitaracsa has been an exception. In 1983 *Tio Pancho* petitioned for the establishment of a full primary and secondary educational curriculum in Quitaracsa, and the Ministry of Education granted the request. While the construction of the school only completed more than a decade later in the 1990s, at least the community now had an educational centre. However, the quality of the curriculum being offered has been subject to debate. In many instances, teachers with troublesome pedagogical records were sent to Quitaracsa to work. Most recently, a teacher with a child abuse record was sent to work in Quitaracsa. Additionally, as recent studies on Indigenous education have shown, students who receive education in rural areas are unable to compete in higher education contexts. And, the majority of the information they receive in class hold Western values and history at a higher pedestal than Indigenous customs, history, or identity.[60] Such is perpetuated further by the strong push for Spanish and majority-culture-customs from the teachers. For instance, even for Quitaracsa that does have a primary and secondary school, the teachers sent there to work still lacked basic knowledge of Quechua. The emphasis in the classroom was and continues to be on turning Quechua-speakers into Spanish-speakers—the so called "intercultural bilingual education." Furthermore, classes focus on propagating Eurocentric knowledge to the students. The ideas are imposed rather than debated along with Indigenous values, customs, and knowledge. Given the societal dynamics of ethnicity and class, students often come out of the educational setting depreciating the culture that shaped their identity. That was and still is a common outcome of the Peruvian educational system, with numerous people with an Indigenous-mother-tongue background denying the identity and pretending not to understand Quechua on the streets. In terms of the conversion of Quechua-speaking to Spanish-speaking individuals, I have seen this first-hand in my most recent visit to Quitaracsa. My cousins, the sons of *Tio Graciano*, were solving math questions with prompts written in a language that they could not understand. So, they could not ask my Quechua-speaking uncle for help when attempting to complete the assignment. As a result, they have been lagging in their studies.

Regardless of the limited impact such educational initiatives had in changing the course of intercultural bilingual education, *Quitaracsinos'* were at least given access to education. My father, *Tio Pancho, Taita Matildo*, and many others had brought this forward. Their travels to Lima, despite the threats they faced, had brought the community an opportunity to have access to education. However, for members of most other communities, the pursuit of education would still require them today to travel to a city, just as it had been the case for my father. This also implied that they had to be "away from family and friends and learn new forms of life," including different culinary habits.[61] Such had been my father's experiences. And, if the individual were so fortunate to accede even higher education opportunities, they would be exposed to "frequent acts of discrimination," which might jeopardize their academics and interpersonal relations. Their current experiences would reflect my father's story, who had to face discrimination before and during the educational process, including primary, secondary, and higher education. His experiences continued being a part of the present for Indigenous communities throughout Peru and shed light as to what will come in the future for Indigenous peoples.

Notes

1 El Peruano, *Normas Legales Año XXI-No. 8437*. Lima: El Peruano, 2003, 248946.
2 Ligia (Licho) López Lopez, *The Making of Indigeneity, Curriculum History, and the Limits of Diversity*. New York: Routledge, 47, 2018.
3 Gerald Vizenor, *Manifest Manners: Narratives on Postindian Survivance*. Lincoln: University of Nebraska Press, vii, 1994.
4 Javicho is the diminutive form for Javier, in reference to my father's first name.
5 Francisco Javier Carranza Romero, "Interview with Francisco Carranza," Personal Interview. Baltimore, July 7, 2019. The stories told in this chapter reflect the oral history provided by Francisco Javier Carranza Romero.
6 Carolina Jorquera, "Sistemas educativos, hegemonía y lógica de conocimiento: Validación, educabilidad y borraduras de la diferencia," in *Claro de Luz: Colonización e Intelectualidades Indígenas en Abya Yala, siglos XX-XXI,* eds. Pedro Canales Tapia and Carmen Rea Campos. Santiago: Instituto de Estudios Avanzados, 149–155.
7 Marta Kania Kania, "The Rights of Indigenous Peoples in Peru: From Socio-Political Marginalization to the Modern Principles of Multiculturalism," *Ad Americam,* 17: 14, 2016.
8 Roger Merino, "La nación reimaginada: Autodeterminación indígena y las olas de indigenismo legal en el Perú," In *Descolonizar el Derecho*, eds. Roger Merino and Areli Valencia. Lima: Palestra Editores, 112, 2018.

9 República del Perú, *Constitución para la República del Perú*. Lima: República del Perú, 4, 1920.
10 Gladys Tzul Tzul, "Forma Comunal de la Resistencia," *Dossier: Revista de la Universidad de México*, 3: 107, 2019.
11 República del Perú, *Ley Orgánica de Educación Pública (No. 9359 promulgada el 1. de Abril de 1941),* Lima: Republica del Perú, 1948.
12 Ibid., 20.
13 Ibid., 20.
14 Congreso de la República, *Constitución Política de la República Peruana Sancionada por el Primer Congreso Constituyente el 12 de Noviembre 1823*. Lima: República del Perú, 1–19, 1823.
15 Congreso de la República, "Reseña Histórica del Congreso," Congreso de la República, 2017. Accessed February 1, 2020. http://www.congreso.gob.pe/participacion/museo/congreso/resena-historica-congreso/.
16 The collective rights of Indigenous peoples included the legal recognition of their collective existence and the inalienable and imprescriptible rights of Indigenous peoples to their communal lands under Article 58 of the 1920 Constitution. See Roger Merino, "La nación reimaginada," 109.
17 Ibid., 110.
18 It is important to note that the 1933 Constitution of Peru was the tenth constitution adopted by the Peruvian Congress. As of this writing, Peru has had twelve constitutions, with the latest one being effective from 1993. See Congreso de la República, *Constitución Política del Perú (29 de Marzo de 1933)*. Lima: República del Perú, 1, 1933.
19 Ñusta Carranza Ko, "Forcibly Sterilized: Peru's Indigenous Women and the Battle for Rights," in *Human Rights as Battlefields: Changing Practices and Contestations*, eds. Marie-Christine Doran, Gabriel Blouin-Genest, and Sylvie Paquerot. New York: Palgrave Macmillan, 160–161, 2018.
20 Marisol De la Cadena, *Indigenous Mestizos: The Politics of Race and Culture in Cuzco, Peru, 1919–1991*. Durham: Duke University Press, 45, 2000.
21 Karen Stote, "The Coercive Sterilization of Aboriginal Women in Canada," *American Indian Culture and Research Journal*, 36: 117, 2012.
22 The author refers to Pancho in the chapter as *Tio Pancho*, given the author's relations with the family.
23 República del Perú, *Decreto Ley No. 21156 Que Reconoce Quechua Como Lengua Oficial de la República*. Lima: República del Perú, 1975.
24 República del Perú, *Ley Orgánica de Educación Pública*, 20.
25 Ibid., 18.
26 Ñusta Carranza Ko, "Comparing the effects of Chinese and Japanese migration experiences on Peruvian national identity," *Journal of Chinese Overseas*, 13: 79, 2017.
27 Ñusta Carranza Ko, "Forcibly sterilized," 161.
28 Alberto Flores Galindo, *Buscando un Inca*. Lima: Editorial Horizonte, 276, 1988.
29 Maria Elena García, *Making Indigenous Citizens: Identity, Development and Multicultural Activism in Peru*. Stanford: Stanford University Press, 66-68, 2005.
30 Ñusta Carranza Ko, "Forcibly sterilized," 149–172; Ñusta Carranza Ko, "Comparing the effects of Chinese and Japanese Migration," 70–93.
31 Marisol De la Cadena, *Indigenous Mestizos,* 220–221.
32 Karen B. Graubart, *With Our Labor and Sweat: Indigenous Women and the Formation of Colonial Society in Peru 1550–1700*. Stanford: Stanford University Press, 40, 2007.
33 Carolina Jorquera, "Sistemas educativos," 153.
34 Francisco Carranza Romero, *El Retorno del Yaynu* (Trujillo: Universidad Nacional de Trujillo, 1985); Francisco Carranza Romero, Magical city of Yaynu, (Seoul: Munhak kwa Chisongsa, 1988).

35 United Nations, "Convention against discrimination in education," United Nations Educational, Scientific and Cultural Organization, 1962, Accessed November 1, 2019. https://treaties.un.org/doc/Publication/UNTS/Volume%20429/volume-429-I-6193-English.pdf.
36 Ibid.
37 United Nations, "International covenant on economic, social and cultural rights," Office of the High Commission on Human Rights, 1966. Accessed November 1, 2019. https://www.ohchr.org/en/professionalinterest/pages/cescr.aspx.
38 International Labor Organization, "C169-Indigenous and tribal peoples convention, 1989 (No. 169)," International Labor Organization, 1989. Accessed November 15, 2019. https://www.ilo.org/dyn/normlex/en/f?p=NORMLEXPUB:12100:0::NO::P12100_ILO_CODE:C169.
39 As per Article 55 of Peru's 1993 Constitution, international law in Peru has an automatic effect once the state becomes party to international commitments.
40 Stéphanie Rosseau and Eduardo Dargent, "The construction of Indigenous language rights in Peru: A language regime approach," *Journal of Politics in Latin America,* 11: 165, 2019.
41 Maria Elena García, *Making Indigenous Citizens*, 78.
42 Sonia García-Segura, "Identidad, Lengua y educación: La Realidad de la Amazonía Peruana," *Revista de Estudios y Experiencias en Educación,* 18: 193–207, 2019.
43 Monika Ludescher, "Instituciones y prácticas coloniales en la Amazonia peruana: pasado y presente," *Indiana* 17/18: 335, 2000/2001.
44 Ibid., 335.
45 López Lopez, Ligia (Licho), "Language, science, and a mission: Another look at pluralism," *Knowledge Cultures* 4 (2): 11–12, 2016.
46 Oscar Espinosa, "Educación Superior para Indígenas de la Amazonía Peruana: Balance y Desafíos," *Antropológica* 39: 103–104, 2017.
47 Andrés Chirino Rivera and Martha Zegarra Leyva. *Educación Indígena en el Perú.* Lima: OPECH, 22, 2004.
48 Ibid., 44–46.
49 Ministerio de Educación, *Resolución Viceministerial No. 154-2017-MINEDU.* Lima: República del Perú, 3–4, 2017.
50 Oscar Espinosa, "Educación superior," 105–106.
51 Virginia Zavala, *Avances y Desafíos de la Educación Intercultural Bilingüe en Bolivia, Ecuador y Perú.* Lima: CARE Perú, 193, 2007.
52 Edwin Cruz Rodríguez, "¿Que tan interculturales han sido las políticas de Educación Intercultural Bilingüe (EIB)? Una reflexión sobre los casos de Bolivia y Ecuador," *Magistro* 8, (15): 64, 2014.
53 AIDESEP (Asociación Interétnica de Desarrollo de la Selva Peruana), "AIDESEP rechaza al Instituto lingüístico de Verano," AIDESEP (Asociación Interétnica de Desarrollo de la Selva Peruana), 2015. Accessed April 2, 2020. http://www.aidesep.org.pe/node/12376.
54 Oscar Espinosa, "Educación superior," 113.
55 Ministerio de Educación, *Resolución Viceministerial*, 6–7.
56 CHIRAPAQ, *Juventud, Educación Superior y Movimiento Indígena en el Perú.* Lima: CHIRAPAQ, 19, 2014.
57 Ibid., 108.
58 Ibid., 111.
59 Ibid., 14.
60 Ibid., 15.
61 Ibid., 14.

References

AIDESEP (Asociación Interétnica de Desarrollo de la Selva Peruana). "AIDESEP rechaza al Instituto lingüístico de Verano." AIDESEP (Asociación Interétnica de Desarrollo de la Selva Peruana), 2015. Accessed April 2, 2020. http://www.aidesep.org.pe/node/12376.

Carranza Ko, Ñusta. "Forcibly sterilized: Peru's Indigenous Women and the Battle for Rights." In *Human Rights as Battlefields: Changing Practices and Contestations*. Marie-Christine Doran, Gabriel Blouin-Genest, and Sylvie Paquerot, eds. New York: Palgrave Macmillan, 149–172, 2018.

CHIRAPAQ. *Juventud, Educación Superior y Movimiento Indígena en el Perú: Resumen y Propuesta*. Lima: CHIRAPAQ, 2014.

Chirino Rivera, Andrés and Martha Zegarra Leyva. *Educación Indígena en el Perú*. Lima: OPECH, 2004.

Congreso de la República. *Constitución Política de la República Peruana Sancionada por el Primer Congreso Constituyente el 12 de Noviembre de 1823*. Lima: República del Perú, 1823.

Congreso de la República. *Constitución Política del Perú (29 de Marzo de 1933)*. Lima: República del Perú, 1933.

De la Cadena, Marisol. *Indigenous Mestizos: The Politics of Race and Culture in Cuzco, Peru, 1919–1991*. Durham: Duke University Press, 2000.

Espinosa, Oscar. "Educación Superior para Indígenas de la Amazonía Peruana: Balance y Desafíos." *Anthropologica* 39: 99–122, 2017.

Flores Galindo, Alberto. *Buscando un Inca*. Lima: Editorial Horizonte, 1988.

García-Segura, Sonía. "Identidad, Lengua y educación: La Realidad de la Amazonía Peruana." *Revista de Estudios y Experiencias en Educación*, 18: 193–207, 2019.

Graubart, Karen B. *With Our Labor and Sweat: Indigenous Women and the Formation of Colonial Society in Peru 1550–1700*. Stanford, CA: Stanford University Press, 2007.

International Labor Organization. "C169-Indigenous and Tribal Peoples Convention, 1989 (No. 169)." International Labor Organization, 1989. Accessed November 15, 2019. https://www.ilo.org/dyn/normlex/en/f?p=NORMLEXPUB:12100:0::NO::P12100_ILO_CODE:C169.

Jorquera, Carolina. "Sistemas educativos, hegemonía y lógica de conocimiento: Validación, educabilidad y borraduras de la diferencia." In *Claro de Luz: Colonización e Intelectualidades Indígenas en Abya Yala, siglos XX-XXI*. Pedro Canales Tapia and Carmen Rea Campos, eds. Santiago: Instituto de Estudios Avanzados, 149–155, 2013.

Kania, Marta Kania. "The rights of Indigenous Peoples in Peru: From socio-political marginalization to the modern principles of multiculturalism." *Ad American*, 1: 11–32, 2016.

López Lopez, Ligia (Licho). "Language, science, and a mission: Another look at pluralism." *Knowledge Cultures* 4, (2): 7–24, 2016.

López Lopez, Ligia (Licho). *The Making of Indigeneity, Curriculum History, and the Limits of Diversity*. New York: Routledge, 2018.

Ludescher, Monika. "Instituciones y prácticas coloniales en la Amazonia peruana: pasado y presente." *Indiana* 17/18: 313–359, 2000/2001.

Merino, Roger. "La nación reimaginada: Autodeterminación indígena y las olas de indigenismo legal en el Perú." In *Descolonizar el Derecho, eds. Roger Merino and Areli Valencia*, 97–130. Lima: Palestra Editores, 2018.

Ministerio de Educación. *Resolución Viceministerial No. 154–2017-MINEDU*. Lima: República del Perú, 2017.

República del Perú. *Constitución para la República del Perú*. Lima: República del Perú, 1920.

República del Perú. *Decreto Ley No. 21156 Que Reconoce Quechua Como Lengua Oficial de la República*. Lima: República del Perú, 1975.

República del Perú. *Ley Orgánica de Educación Pública (No. 9359 promulgada el 1. De Abril de 1941)*. Lima: República del Perú, 1948.

Rodríguez, Edwin Cruz. "¿Que tan interculturales han sido las políticas de Educación Intercultural Bilingüe (EIB)? Una reflexión sobre los casos de Bolivia y Ecuador." *Magistro* 8, (15): 59–87, 2014.

Rosseau, Stéphanie and Eduardo Dargent. "The construction of Indigenous language rights in Peru: A language regime approach." *Journal of Politics in Latin America*, 11: 161–180, 2019.

Tzul Tzul, Gladys. "Forma Comunal de la Resistencia." *Dossier: Revista de la Universidad de México*, 3: 105–111, 2019.

United Nations. "Convention against discrimination in education." United Nations Educational, Scientific and Cultural Organization, 1962. Accessed November 1, 2019. https://treaties.un.org/doc/Publication/UNTS/Volume%20429/volume-429-I-6193-English.pdf.

United Nations. "International covenant on economic, social and cultural rights." Office of the High Commission on Human Rights, 1966. Accessed November 1, 2019. https://www.ohchr.org/en/professionalinterest/pages/cescr.aspx.

Vizenor, Gerald. *Manifest Manners: Narratives on Postindian Survivance*. Lincoln, NE: University of Nebraska Press, 1994.

Zavala, Virginia. *Avances y Desafíos de la Educación Intercultural Bilingüe en Bolivia, Ecuador y Perú*. Lima: CARE Peru, 2007.

7 Preparing Teachers Through Land Education

Indigenous Erasure, Reclamation, and Resurgence in Campus Spaces

Jan Hare, Christine Bridge and Amber Shilling

Introduction

Indigenous ways of knowing emerge from Indigenous peoples' relationship to land and place. Knowledge of, experiences with, and relationship to land and place are expressed through languages, stories, histories, ceremonies, and practices rooted in Indigenous peoples' long inhabitation of a particular place.[1] Indigenous perspectives on land and place emphasize connections between human and more-than-human worlds, recognize the centrality of land/place to ancestral languages, and engage with themes of resistance and resurgence in response to colonialism both past and present. These critical Indigenous perspectives tend to be overlooked in place-based, ecological, outdoor, and environmental education approaches used in teaching and learning with pre-service teachers, ignoring Indigenous priorities of decolonization, sovereignty, and reconciliation.

Approaches in teacher education that advance place-based learning or eco-pedagogies as part of curriculum tend to foreground stewardship for the environment, outdoor education and play, gaining a deeper understanding of local and global environments, and competence in outdoor spaces.[2] Others caution against appropriation when Indigenous practices are made part of physical education, describing how stereotypes of Indigenous peoples, histories, and cultures can be played out in settler myths, such as being at "one with nature" or escaping nature in order to "go Native."[3] In fact, "the overly deterministic environmentalism that marks many outdoor and environmental education programs runs squarely up against Indigenous traditions of knowing."[4] Further, knowledge and experiences of land cannot be relegated to a specific curricula, associated only with physical education or Indigenous education coursework in teacher education programs. This poses limitations on the relevance of Indigenous perspectives across teacher education. What is needed within the curriculum of teacher education are opportunities for pre-service teachers to challenge dominant settler-discourses and experiences that help them rethink their relationship to the land and place in ways that draw from Indigenous knowledge traditions, situated within local Indigenous context, and are concerned with Indigenous priorities.[5]

In this chapter, teacher-researchers explore pre- and in-service teachers' reflections on a post-secondary institution's colonial relationship to Indigenous histories and lands, using the physical campus as a means to confront settler-colonialism in educational spaces through an emphasis on land education.[6] We report on the ways largely settler-dominant teachers engage with the colonial legacies of higher education spaces as sites for potential inquiry and their responses to the resurgence and reclamation of Indigenous knowledges, stories, and histories buried under past and present campus land developments. Through experiential learning activities focussed on learning with, on, and from land and place, we see the potential of teachers to denaturalize their understandings of colonial arrangements by empowering ancient and dynamic knowledge resurfacing on campus spaces.[7] Teachers then engage in decolonizing dialogues, asking them to respond to narratives and counternarratives to critically assess erasure and presence of Indigenous people, communities, and knowledges.

Approaches to prepare pre- and in-service teachers to engage with Indigenous knowledges and communities requires them to engage with Indigenous lands.[8] Public post-secondary institutions occupy Indigenous lands and territories. It is through claiming title to Indigenous lands that post-secondary institutions were established, generate revenue and resources, and drive knowledge, research, and learning agendas. Processes of reconciliation, decolonization, and self-determination require higher learning institutions to reset their relationships with Indigenous peoples on whose lands Indigenous and non-Indigenous students learn and live. Evidence of this can be seen in the growing number of symbolic and material gestures on campus spaces mobilized by Canada's Truth and Reconciliation Commission[9], including new buildings and structures, art work, cultural installations, and teaching and learning spaces. Decolonizing approaches that seek to understand and analyse the physical and material aspects of higher educational settings serve not only to expose colonial practices, but also reveal possibilities for much needed social transformation as part of a post-secondary education.[10]

Our intention is to engage pre- and in-service teachers in re-interpreting the physical and cultural landscapes of the post-secondary campus situated on Musqueam traditional, ancestral, and unceded lands, a place that has been appropriated by the settler city known as Vancouver, Canada, and a place that is familiar to them through their travels to campus for coursework and study. The Musquem hən̓q̓əmin̓əm̓ speaking people have lived on these lands as long as there has been land to live on. We use material, visual, and symbolic examples on campus that demonstrate the vitality and spirit of Indigenous presence for a guided land learning tour we created for the students in a teacher education course activity. Though two of the three sites we describe in this paper focus on Musqueam history, story, and memory, the learning tour is a guided excursion around the university campus to a number of sites that reflect the growing and diverse presence of Indigenous peoples from beyond the local, as well as convey the complexities of Indigenous-settler relations. While we want students taking part in the guided learning tour to grapple with "the material manifestations of collective trauma"[11] that reveal institutions' on-going complicity in colonialism through landmarks, symbols, representations, structures, and Eurocentric knowledge traditions, we also want to demonstrate the ways in which Indigenous knowledge traditions can reshape institutional narratives by attending to voice, space, and agency of Indigenous people and communities. We concur with Metcalfe that "By addressing difficult knowledge of the university at the university, campus/community/classroom are intertwined so as to potentially form new networks for knowledge production and social understanding."[12]

Campus spaces as critical sites of inquiry

Post-secondary institutions are deeply implicated in colonial practices that shape what is valued in curriculum, knowledge production, and cultural representations. This has resulted in the erasure of Indigenous presence within physical campus spaces, marginalization of Indigenous students from higher learning, and the delegitimizing of Indigenous knowledges in classrooms. For example, Minthorn and Nelson examine the macro-structural aspects of a campus college, including physical spaces of the dormitory, library, campus buildings, signs, and plaques that are embedded with racist and genocidal values, albeit commemorated during campus tours aimed at recruiting and promoting the institution.[13] Responses to the on-going vestiges of overt colonialism on the campus are also being played out in the news and social media. In 2018, Ryerson University, located in central Canada, faced a student-led petition to remove a statue of Egerton Ryerson due to his role as one of the "architects" of the Canadian Indian Residential School system, a colonial form of schooling that forced generations of Indigenous children to be removed from their families and communities to attend state designed institutions.[14] The purpose of these schools was to assimilate and eliminate the "Indian from the child." The impact of this form of schooling has been cultural genocide, and the intergenerational legacy it has left behind is responsible for conditions endemic in many Indigenous communities today. Instead of removing the statue as requested, the school installed a plaque nearby that acknowledges Ryerson's role in the cultural genocide perpetuated by the Residential School system. In another example at an eastern Canadian post-secondary institution, a McGill University anti-colonial student group vandalized the statues of Queen Victoria and James McGill "signalling contempt for the figures who represent imperialist and racist histories".[15] These colonial artifacts, highly visible on campus spaces, form a "colonial wall" designed to confine, control, or contain dominant narratives of the settling of a nation.[16]

These examples of student-driven resistance at higher education institutions can be viewed in light of Canada's Truth and Reconciliation Commission's 94 Calls-to-Action, a set of recommendations that includes the call for education institutions to commit to educating students about the history of the genocide of Indigenous people in Canada. Stein points out that much of the work of identifying colonial founding structures in post-secondary institutions occurred after the TRC's Calls to Action were published.[17] This also coincides with the timing of many institutions' rhetorical commitments to reconciliation and Indigenization.[18] For Stein, truth telling and critical thinking with respect to a post-secondary institution's accurate reflection of colonial legacy and its impacts, are a guide to meaningful action forward.[19]

Over time, the university where we work has been increasing Indigenous presence by making Indigenous visual expressions, representations and languages more prominent, thus giving significance to Indigenous perspectives and community relationships. Like other higher education institutions, the university seeks to acknowledge the pre-colonial legacy of the land it occupies, reveal Indigenous histories erased through development, and reconcile its relationships with local Indigenous communities. What we are currently seeing on campuses is what anthropologist Keith Basso calls "place making"[20], a retrospective world-building where the land and each place in the landscape "conjures up everything that ever happened there, and make it present again in the community."[21] Examples of this place-making can be seen in initiatives, such as the Land Beneath Our Feet at the University of British Columbia, Indigenous Placemaking at Ryerson University, Memory Keepers at Concordia University, the Coyote Project at Thompson River University, Cyclical Motion at the University of Manitoba, and the Indigenous Landscape Project at the University of Toronto.

Indigenous lifeworlds on the physical landscapes and in living arrangements, study places, classrooms, or social spaces, create welcoming environments for Indigenous students and community members who have long been marginalized in academia. As post-secondary institutions try to rebuild their relationships with local Indigenous communities, there are increasing efforts to introduce students, faculty, staff, and visitors to the rich land and place-based histories of Indigenous communities. In planning our guided land learning tour, we recognize that not only do the lands and places we select hold meaning, we also believe they hold spirit. This spirit takes its form in the intentions and materials used by artists, designers, and knowledge keepers in their creations that that artists, designers, and knowledge keepers use in their creations that become part of campus spaces. The meaning and stories embedded within the learning objects or symbols are cultural bridges to ancestors and present- day relatives, and within protocols associated with construction, configuration, and location of Indigenous presence that might include consultation with Indigenous people and community, ceremony, and protections of Indigenous knowledge. We propose that spirit in place-making is about connectedness to Indigenous lifeworlds and considers the sacred, where landscape, sculptures, objects, living things are connected to more-than-human worlds, are sources of energy, uphold Indigenous knowledge and practices, and remain significant to Indigenous people.

Land education

Land is central to any discussion regarding decolonization, reconciliation, and self-determination. For our purposes, the concept of land refers to more than physical geographic space. It encompasses all water, earth, air, spirit, history, and cosmology that reside within specific spaces. Similar to other scholars, we view land as a living and sentient being, and our discourses and interactions with the land are shaped and informed by our vision, pedagogies, teaching practices, and meaning-making.[22] Land is a starting point, both geographically and epistemologically, from which educators can draw awareness to dominant discourses and challenge embedded issues of colonialism. Land also provides a pedagogical frame and respectful space that presents the opportunity for educators to explore Indigenous epistemologies and look at ways in which they might be integrated into their teaching practice. We use land education as a theoretical framework to shape the guided learning tour, activities, and discussion dialogues, as well as interpret pre- and in-service teachers' responses to the learning tour. This is because land education encompasses specific geographies that hold physical, spiritual, and linguistic dimensions of Indigenous peoples' long inhabitation of a place.[23] Land education is described by Tuck, McKenzie, and McCoy as a pedagogy that situates Indigenous epistemological and ontological accounts of land at the centre of discussion.[24] This includes Indigenous understandings of land, Indigenous language in relation to land, and Indigenous critiques of settler colonialism.

In particular, we are drawn to Calderon's description of land-based education for several reasons.[25] For Calderon, land education problematizes the relationship between land and settler colonialism, uncovering how settler colonial projects are maintained and produced. She asserts that land education takes into account Indigenous rights and sovereignty, and environmental and ecological sensibilities. As such, it serves to disrupt settler identities. Particularly helpful to our work is Calderon's consideration of reinhabitation as part of land education. Reinhabitation "occurs when local, democratic self-management of degraded homelands becomes possible and stakeholders come to understand the colonizing effects of past historical practices."[26] She argues that issues relating to sustainability and the land cannot be understood if Indigenous communities are not central.

Calderon writes that "a land education model demands we decolonize the 'local' in order to understand how settler colonialism is currently enacted and taught."[27] As a pedagogical frame, land education provides a starting point from which educators are able to learn Indigenous epistemologies, and facilitate discussion regarding colonialism and dominant discourses surrounding land and land use. Hence, the learning tour designed for our pre- and in-service teachers was based on this model. Our goal was to engage learners in an experiential activity that would take them to local campus places reflective of Indigenous presence, that spoke to erasure and displacement of Indigenous people in the local area, and that also drew attention to re-inscription of that presence.

Participants were provided with guiding questions to help them consider and capture the various layers of stories and "colonial fill"[28] present in each campus place, with the intent of challenging and disrupting preconceived notions, while at the same time, building new knowledge and social understanding.

Land education learning tour

We teach and research at the University of British Columbia (UBC), a large research-intensive university in Western Canada, that occupies the traditional, ancestral, and unceded territory of the Musqueam people. Two of us identify as Indigenous scholar-educators and one of us identifies as a settler scholar-educator. We teach pre-service and in-service teachers and designed this guided learning tour in the context of a foundational Indigenous education course that all teacher candidates take as part of their professional preparation. Some aspects of the learning tour were extended and adapted to doctoral research conducted by our settler scholar-educator co-author.[29]

Decolonizing goals of this foundational Indigenous education course include developing knowledge of colonial histories and contemporary expressions, as well as providing pre-service teachers with knowledge and a deeper understanding of Indigenous peoples' world views, pedagogies, histories, and contemporary realities. As part of the curriculum, students explore how Indigenous perspectives and worldviews can be incorporated into teaching and learning. In our planning for the course, we designed a guided learning tour, whereby we walked with students, stopping at specific sites that consider not only the physical aspects of the place, but also the historical, emotional, and sociopolitical contexts that create and inform the students' experiences of the site.[30] Similar to Minthorn and Nelson[31], who analyze how colonialism operates in the physical sites of campus spaces, we see the sites of exploration as more than a collection of inanimate objects. We view these cultural artifacts as holding energies that are constantly interacting; a force to the human experience.[32]

Within the context of research, Bridge conducted a study that included elements of the learning tour.[33] The central question that guided her study focused on the ways in which land-based pedagogies can become a framework through which educators can integrate land education into their practice. This research involved in-service educators participating in a series of land-based activities at various public sites on or near the university campus. Participants in her study reflected on a series of guiding questions and data was collected from reflective journals and semi-structured interviews. Some of the sites included in her study derived from, and became a complement to, the guided learning tour we share in this chapter.

Sites of inquiry and student responses

We describe three sites of inquiry that formed a small part of the guided land learning tour organized around experiential learning. As students walked to each campus place, as course instructors we shared histories and stories of the artifacts, structures, and context for the site. At each site, students were encouraged to ask questions, share observations, draw pictures, observe the place site from multiple vantages, take notes, and engage in critical conversations with the instructor and with each other. A set of guiding questions specific to each site was provided and students were given the opportunity to reflect on their experiences prior to providing written responses. Questions prompted students to consider how these sites of inquiry have been materially, spatially, symbolically, and ideologically constructed or positioned to erase or appropriate Indigenous knowledges, histories, and presence. In addition, they were asked to reflect on how each site reflected themes of Indigenous reclamation, resurgence, and reconciliation that had been discussed in coursework.

Musqueam street signs

One of the sites focussed on street signs installed on various locations around the university campus created in partnership with the Musqueam First Nation to include hən̓q̓əmin̓əm̓ and English. Nine of the 54 signs make visible Musqueam presence by juxtaposing the languages against one another on the signs, demonstrating linguistic differences and world views. The English names refer to cardinal directions of north, south, east, west, and central places on campus, such as Main Mall or Student Union Boulevard. Musqueam language draws from the geography and uses a place-based orientation towards the land or water, using concepts such as "up the river," "inland towards the shore," or "middle" to signal Main Mall. The names chosen by Musqueam "seek to educate us about the way they perceive place, movement across the land, and to show everyone how their language and culture is intrinsically connected to their territory."[34]

Figure 7.1 Paul H. Joseph / UBC Brand & Marketing

For the Musqueam, land and living are intimately connected. History, made available to the public through their community website, explains this connection:

> The one called χe:l̓s (the transformer) arrived. He took pity on the people and began travelling alongside them to help. The people and creatures that were not right were taught. χe:l̓s taught us empathy, charity, forgiveness, compassion, and the importance of sharing the land, as well as other responsibilities. He left his teachings where he walked, and these teachings grew into our law. Those who refused to learn, χe:l̓s fixed, transforming them. Many were turned to stone and others into animals. These travels and transformations are written in the earth, captured in our sχʷəy̓em̓ (ancient histories), and recorded in our place names, making these lands core to our teachings. To this day, this ancestral knowledge and our connections to these places are reaffirmed through our cultural teachings, practices, and responsibilities. These laws are the foundations of our present and future success.[35]

Guiding questions and student responses

There were three questions that students reflected on and responded to following their visit to the street signs:

- How are perspectives of land and place reflected within Indigenous languages?
- How do Indigenous perspectives on the relationship between land and language diverge from Western/Eurocentric perspectives?
- How do space and place converge (and diverge) with Indigenous priorities?

After visiting, discussing, and reflecting on the Musqueam street signs, students expressed greater understanding of the links between Indigenous landscapes, land, culture, and language. As one student wrote, "*The land is a source of knowledge and it is therefore also linked with the language. It is not enough for us to give land acknowledgements. We must also honour the language of the Indigenous people in the area we are in.*"

When comparing the place-based directional system of the Musqueam street signs to the English counterpart names, many students were struck by the divergence of perspectives regarding one's relationship to and with the land, and how the signs were reflective of these perspectives. As one student noted: "*the focus of Western perspectives of land revolve around property, ownership, and the notion that the Western world discovered this land. In contrast, Indigenous perspectives focus heavily on connection with the earth, the land and with nature.*"

Students also commented on perspectives and relationships between place and space. For example, one student wrote, "*Indigenous communities make meaning from the land that they have historically occupied for thousands of years. Since colonialism, these 'spaces' have been systematically removed from them and occupied by white settlers. In contemporary politics, place and space are converging as Indigenous people reclaim that space. A space lacks meaning or context, but all of this land has deep cultural significance, therefore Canada (Turtle Island) is indeed an Indigenous place."*

As Styres reminds us, land holds a depth of knowledge and storied relationships that are "etched into the essence of every rock, tree, seed, animal, pathway, and waterway in relation to the Aboriginal people who have existed on the land since time immemorial."[36] Given the time to observe and critically reflect on the street signs was described as a "gift" by several students. Similar to another site described below on the guided land learning tour, students expressed regret that they had walked by the street signs hundreds of times without ever pausing to consider the layers of meaning and perspective held in each specific space and place. They were quick to acknowledge the importance of their role as educators in building student awareness of, and multiple understandings about, the local places within which they live and work.

Native hosts

Native Hosts is a permanent twelve-piece sign series created by Cheyenne and Arapaho artist and scholar, Hock E Aye Vi Edgar Heap of Birds. These white aluminum signs are situated around the University of British Columbia campus in high volume pedestrian zones with the purpose of disrupting the recognizable norm of road signs, positioning non-Indigenous viewers as guests on First Nations lands, lands inhabited prior to colonization by named Indigenous hosts. Viewers can give pause and reflect on the twelve traditional territories throughout British Columbia represented in the works. Heap of Birds' *Native Hosts* signs can also be viewed internationally, as part of exhibits or permanent installations elsewhere, where they serve as institutional place markers of Indigenous lands. The settler name of the place on the sign is backwards, suggesting that the state's colonial violence against Indigenous people warrants critical treatment.[37] According to Heap of Birds, he flips the name of the province (or state) as a way of displacing its authority: "It's about asserting sovereignty, and calling yourself or a tribe a 'host' is a gentle way of asserting ownership."[38] Within the guided land learning tour, these land markers became a focal point for prompting pre- and in-service teachers to think about colonial histories that attend to specific places, to consider the ways in which these material gestures can intervene on our understanding of land as more than physical space, and also to consider Indigenous priorities concerned with land rights, treaties, and sovereignty.

Guiding questions and student responses

These were the questions that students reflected on and responded to following their visit to the Native Hosts signs:

- What knowledge and practices of Indigenous people become lost or buried under University of British Columbia campus developments?
- How do these material and symbolic representations reflecting Indigenous-settler relations shape your understanding of land and place?
- How are the material, symbolic, spiritual, and ideological understandings of land conveyed at this site detached from Eurocentric understandings of land and place (e.g. property, ownership, sacredness, relationship, responsibility, resources)?
- What new understandings emerge for you at this site that challenge your thinking about land and place?

Calderon notes that land education models demand decolonizing the "local" in order to see how settler colonialism is enacted and taught.[39] Learner responses to the Head of Birds' installations brought this into sharp focus. Students commented on the sense of immediacy and disruption that was evoked by the signs. For example, a student noted that of all the sites visited during the learning tour, the Native Hosts installations challenged her thinking: "*This was the place that shaped my understanding the most. This is because of the word 'today.' Today, and every day, we, the institution, are being hosted.*"

This sense of disruption was echoed by others. Another student shared this tension that informed their understanding, "*the backwards British Columbia disrupts authority as BC then becomes an Idea rather than a Fact. It makes us think about the fact that we are using the land, maybe not as it was meant to be used. We are just guests to this place. It is not ours.*"

The Heap of Birds' installations also demonstrated the limitations in their awareness of the routines of coloniality. As one student commented: *"I walk by [the sign] three days a week and I didn't see it until I did this activity. I just walked right past it – didn't see it.... It speaks to me – maybe of something that is not directly related, but the goal of the art piece. It makes me think of how we live with Indigenous people, and we are on their land that we stole, and we walk right past without even noticing."*

The fact that the sign installation had gone unnoticed by many students prior to the learning tour was juxtaposed to the impact of its message: *"the art feels like a form of resistance. It's on us to make sure we do not forget. Don't forget the land you are standing on and the people that it belongs to. Sometimes, it will feel easy to forget and a lot of people would very much prefer it if you did, but don't. Resist and keep looking. That's the only way reconciliation is possible."*

McGinty and Bang remind us that a fundamental view of settler-colonial societies is "the acquisition of land as property and a logic of Indigenous elimination, followed by the establishment of settler life ways as the normative benchmark from which to measure development."[40] This site, in particular, precipitated student responses that demonstrate the complexities of decolonial processes fraught with frustration and limitations in learning about land dispossession that occurs through colonial policies and the complicity of post-secondary institutions in settler colonialism. Correspondingly, the installation demonstrates the "interrupted and interruptive capacity"[41] of Indigenous perspectives on sovereignty and Indigenous-settler relations.

The Musqueam welcome post

As previously mentioned, our campus is situated on Musqueam traditional, ancestral, and unceded lands. Although the university was established on these lands in 1908, the Musqueam people have lived in this location for thousands of years. As the University began clearing the land, building, and expanding, a public history of UBC began to unfold, overwriting Musqueam's much longer history and presence.[42] Like other cultural, educational, and civic institutions, the University became "heavily invested in imagined representations of Indigenous people and their cultural practices, which manifested in different ways in its operations and public spaces."[43] This is observed in the number of Northwest Coast house posts displayed on campus.

House posts are a form of a totem pole. While a house post serves an architectural function in the communal and cultural space of West Coast longhouse structures, they are understood as monuments raised to acknowledge the ancestors or honour a person. They may also be declarations of one's family or histories and hold significant meaning. The 34-foot Musqueam welcoming pole, carved by Musqueam artist Brent Sparrow Jr., was presented as a gift to the University as part of its 100th year celebration. The post's size and magnificence do not go unnoticed in the busy pedestrian area where it is located at *šxʷʔeýeʔ* (middle).

Also known as the Musqueam sʔi:łqəý qeqən (double-headed serpent post), the post tells the story behind the Musqueam's traditional name, xʷməθkʷəy̓əm. A long time ago, a double-headed serpent (sʔi:łqəý) fertilized the area, causing a unique plant, məθkʷəý, (River Grass) to flourish. As a result, the people living in the area became known as xʷməθkʷəy̓əm (Musqueam—People of the River Grass).[44] The post is part of a larger installation that includes an elegantly tiered water flow to symbolically capture the image of a river. River grass grows in abundance along the route of the "river" and around the post. The post serves as a permanent welcome to all visitors to campus, and as a reminder of the relationship with the Musqueam people who were here long before the university's history began. This site of inquiry allowed for a focus on the institutional relationships with Indigenous people and how land, as a source of knowledge, is steeped in history and memory.

Guiding questions and student responses

The guiding questions used in conjunction with the visit to the Musqueam welcome post were the following:

- How does land serve as a source of knowledge? What authority do you give to that knowledge?

- What stories emerge from land and place?
- How do these material and symbolic aspects of culture at this place shape your understanding of relationships to local Indigenous lands and communities?
- What is the potential of the welcoming post in shaping Indigenous-settler relations?

In documenting reactions to visiting the Musqueam welcome post, students expressed reverence for the beauty of the post, as well as for the story it conveyed. As one student commented: *"The Musqueam Post is striking – a presence of art and expression of identity that reminds us daily to respect the First People that were here before us. It helps deepen our understanding about Musqueam stories, culture, and connections to this land."*

Many students focussed on connections between land and identity in their responses. *"I see how land and place are so intimately connected to identity for Indigenous people. Given this, I have a deeper understanding of Indigenous people's motivation and perseverance in reclaiming their presence on their land, and the importance of settlers learning about this as an aspect of reconciliation."*

Similar to reactions to the other sites of inquiry, students gave pause to consider the implications of land and loss of land due to campus development. Themes expressed in the following response were echoed in others: *"When [the university] develops the land, Indigenous history and language are lost and buried. The stories are connected to the land, and the language is connected to place and the stories of the land. The appropriation of this land for exploitation and gain disrupts the whole Indigenous way of being and destroys the harmony by changing the landscape."*

The story of the welcome post and the origin of the Musqueam name are emblematic of how Indigenous knowledge is inherently tied to land, where meaning and identity are constructed through landscapes, territory, and land formations.[45] Land is considered a source of knowledge and authority, where meaning is formed through relationships and experiences.[46]

Pedagogical practices for land education on campus spaces

Our goal as teacher educators and researchers was to challenge dominant discourses with respect to land and place, and use land education and place-making to shape our pedagogical practices and land learning tour. In this chapter, our focus was on the visible and material aspect of the campus space that included street signs, an art installation, and a welcome pole. Interventions being undertaken at other institutions that affirm Indigenous identity and presence can be observed in land acknowledgements at public gatherings; Indigenous-centered academic, social, and cultural/ceremonial spaces; promotion of Indigenous languages on signage; commemorative plaques; planting of traditional flora and medicines; micro-landscapes that reflect Indigenous ecologies; Indigenous community flags; renaming sports teams and redesigning symbols; and other material artifacts of Indigenous cultures. These physical, material, and symbolic additions to campuses may highlight commitments to reconciliation, but we need to attend to how these interventions on campuses are connecting students, staff, faculty, and visitors to the land.

We offer pedagogical considerations for those teaching on Indigenous lands who want to create learning experiences for students in their coursework, and to institutions who seek to develop cross-cultural learning opportunities for those who live, study, and visit. Pedagogical practices should give focus to preparation, decolonization, and learning from Indigenous people and perspectives.

Preparation

Higher education institutions occupy Indigenous lands. Instructors and students should have knowledge of local Indigenous people and their histories in the places where they live, learn, and work prior to engaging in land activities or discussions. Students should also be provided with information about campus sites prior to self- or guided explorations, ensuring a critical review of the information acquired for sharing with students. It is possible that students may experience denial or feelings of guilt and shame as they engage with shared colonial histories in which they are implicated. Therefore, it is equally important that students explore their own histories in relation to Indigenous people. Critical pedagogies that allow students to critique their own ways of knowing facilitate opportunities to interrogate their knowledge systems, ideologies, and societal structures.[47]

Decolonization

The physical landscapes of post-secondary can no longer be narrated through discourses of occupation, dispossession, and erasure, leaving modest space and place for Indigenous people's priorities. It is critical for faculty, staff, and students to see how higher education is implicated in colonialism and on-going investments in colonial habits that come from occupation of Indigenous lands. But simply learning about colonial relations is not sufficient for decolonizing land-education approaches. Learning from Indigenous peoples' histories, world views, practices, perspectives, and experiences, as well as experiential approaches, are required to provoke critical thinking and challenge long-held assumptions about the places where we study and work.

Learning on, with, and from land where students attend post-secondary school should move beyond teaching about Indigenous people, their cultures and histories. Teaching should avoid focusing on themes that romanticize Indigenous peoples' relationship with the land or give primacy to themes of stewardship or sustainability. Rather, students should be guided through the process of decolonization by making explicit how colonial projects and settler narratives are constructed, maintained, and reproduced as they concern land and place.[48] Deepening and challenging student assumptions about land and place can occur through critical dialogue. Alongside the questions we have suggested in our guided tour, Stein poses some helpful questions that instructors might consider in their land education approaches:

- By what processes did the land on which the institution sits come to be held by the institution? Who (including humans and other-than-human beings) were affected by those processes?[49]
- Is there a relationship between the institution and local Indigenous community/com-munities, and if so, how is power distributed within that relationship?[50]
- Can I surrender my desires for control, authority, certainty, and security so that I might develop the humility and stamina that are required to be a part of a long-term, multi-layered, messy transformation toward decolonial futures that are not-yet-knowable?[51]

Engaging with Indigenous People and Perspectives for Indigenous Futures

While public spaces and places on campus offer accessible and engaging ways to have complex conversations about Indigenous histories, rights, and contemporary realities, activating Indigenous presence throughout campus landscapes also has implications for Indigenous futures. Transformation of public higher education institutions can only occur when led by Indigenous people and where Indigenous knowledges, bundled in Indigenous histories, traditions, and practices, are woven with the present to create Indigenous futures. Changing the way material and physical environments are organized through Indigenous knowledges speaks back against colonial logics and creates opportunities to imagine land from another perspective. To imagine a post-secondary education beyond settler-colonial manifestations requires Indigenous contributions.

Instructors might consider:

- Complementing Indigenous sites of inquiry with Indigenous political, historical, literary, and community voices.
- Using technology (e.g., apps or links) to assist poems, songs, or images to be accompanied on learning?
- Attend to the broader context that might include Indigenous flora, sounds, stories, and Indigenous teachings that shape learning?
- Ask students how sites of inquiry are linked to Indigenous priorities and futures?

Given that academics are responsible for teaching and learning, they should be establishing learning partnerships with Indigenous community members and scholars to co-construct curriculum concerned with land education. Indigenous communities play a vital role in making visible Indigenous knowledges, histories, and stories on campus and can be partners in teaching and learning. Partnership approaches are effective when instructors are hesitant to engage with Indigenous knowledge for fear of making mistakes, appropriation, or uncertainty of how to incorporate Indigeneity. Teaching and learning that engages Indigenous people in meaningful ways requires humility on the part of instructors. Stein reminds us that the work of transforming higher education must be collective, whereby instructors must be ready to surrender control and authority and carry the pedagogical labour.[52]

Conclusion

The majority of pre- and in-service teachers who participated in this guided land learning tour expressed a commitment to include land-based education as a core part of their teaching practice. Though this meant learning about Indigenous lands and communities, as well as stories tied to land and place, their responses suggest that they still need experiences that unsettle their investments in colonialism. Located on Indigenous lands, higher education institutions create opportunities to engage in critical thinking around colonialism and the physical campus. Place-making is shaping the growing number of Indigenous strategic plans in Canadian higher education and altering the physical landscapes. However, a pedagogical shift towards land education is required if these plans are to realize reconciliation, decolonization, and Indigenous people's self-determination.

Notes

1 Ray Barnhardt and Anagayuqaq Kawagley, "Indigenous knowledge systems and Alaska Native ways of knowing," *Anthropology & Education Quarterly* 36 (1): 9, 2005.
2 Alan Ewart and Jim Sibthorp, *Outdoor Adventure Education: Foundations, Theory, and Research*. Champlaign, IL: Human Kinetics, 2014.
3 Phillip Mullins, Gregory Lowan-Trudeau, and Karen Fox, "Healing the split head of outdoor recreation and outdoor education," in *Routledge International Handbook of Outdoor Studies*.Barbara Humberstone et al, eds. Abingdon, UK: Routledge, 49–58, 2016.
4 Tracy Friedel, "Looking for learning in all the wrong places: Urban native youths' cultured response to western-oriented place-based learning," *International Journal of Qualitative Studies in Education* 24, (5): 540, 2011.
5 Sandra Styres, Celia Haig-Brown, and Melissa Blimkie, "Toward a pedagogy of land: The urban context," *Canadian Journal of Education* 36 (2): 188–221, 2013. Eve Tuck, Marcia McKenzie, and Kate McCoy, "Land education: Indigenous, post-colonial, and decolonizing perspectives on place and environmental education research," *Environmental Education Research* 20, (1): 1–23, 2014.
6 Delores Calderon, "Speaking back to manifest destinies: A land education-based approach to critical curriculum inquiry," *Environmental Education Research* 20, (1), 2014.
7 Cindy Holmes, Sarah Hunt, and Amy Piedalue, "Violence, colonialism, and space: Towards a decolonizing dialogue," *ACME: An International Journal for Critical Geographies* 14, (2): 540–570, 2015.
8 Alexa Scully, "Unsettling place-based education: Whiteness and land in Indigenous education in Canadian teacher education," *Canadian Journal of Native Education* 1 (38), 2015.

9 Truth and Reconciliation Commission of Canada, *Truth and Reconciliation Commission of Canada: Calls to Action,* last modified January 15, 2020, http://trc.ca/assets/pdf/Calls_to_Action_English2.pdf
10 Linda Smith, "Foreword: Keeping a decolonising agenda to the forefront," in *Decolonisation in Aotearoa: Education, Research and Practice.* Jessica Hutchings and Jenny Lee-Morgan, eds. Wellington, NZ: NZCER Press, ix-x, 2016.
11 Amy Metcalfe, "What is difficult knowledge?," *The Difficult Knowledge Project* (blog), last modified January 12, 2020, http://blogs.ubc.ca/difficultknowledge/.
12 Metcalfe, *Difficult Knowledge*, p. 1.
13 Robin Minthorn and Christine Nelson, "Colonized and racist Indigenous campus tour," *Journal of Critical Scholarship on Higher Education and Student Affairs* 4: 73–88, 2018.
14 Raneem Alozzi, "Ryerson unveils plaque revealing its racist history," last modified January 15, 2020 https://theeyeopener.com/2018/06/ryerson-unveils-plaque-recognizing-its-racist-history/.
15 Favour Daka, "Nothing is set in stone: Colonial statues on campus," *The McGill Tribune*, last modified January 15, 2020. http://www.mcgilltribune.com/opinion/nothing-is-set-in-stone-colonial-statues-on-campus-1004/
16 Brenda St. Germain, "Behind the colonial wall: The chains that bind resistance" (Master's thesis, University of Victoria, 2014).
17 Sharon Stein, "'Truth before reconciliation': The difficulties of transforming higher education in settler colonial contexts," *Higher Education Research & Development* 39 (1): 156–170, 2020.
18 Adam Gaudry and Danielle Lorenz, "Indigenization as inclusion, reconciliation, and decolonization: Navigating the different visions for Indigenizing the Canadian Academy," *AlterNative: An International Journal of Indigenous Peoples* 14, (3), 2018.
19 Stein, "Truth before reconciliation."
20 Keith Basso, *Wisdom Sits in Places: Landscape and Language among the Western Apache.* Albuquerque: University of New Mexico Press, 7, 1996.
21 Burton-Christie, , 349, 2009.
22 See Basso, *Wisdom Sits in Places.* Jan Hare, "To know papers: Aboriginal perspectives on literacy," in *Portraits of Literacy Across Families, Communities and Schools.* Jim Anderson et al, eds. Mahwah, NJ: Lawrence Erlbaum Associates, 2005; Styres, Haig-Brown, and Blimkie, "Toward a pedagogy of land".
23 Barnhardt and Kawagley, *Indigenous Knowledge*, 8–10.
24 Eve Tuck, Marcia McKenzie and Kate McCoy, "Land education: Indigenous, post-colonial, and decolonizing perspectives on place and environmental education research," *Environmental Education Research* 20, (1), 2014.
25 Calderon, "Speaking back to manifest destinies."
26 Ibid, 27.
27 Ibid, 28.
28 Metcalfe, *Difficult Knowledge*, p. 2.
29 Christine Bridge, "Land education and reconciliation: Exploring educators' practice." PhD diss., University of British Columbia, 2018.
30 Vine Deloria, Jr. and Daniel Wildcat, *Power and Place.* Golden, CO: Fulcrum Publishing, 2001.
31 Robin Minthorn and Christine Nelson, "Colonized and racist Indigenous campus tour".
32 Ibid, 76.
33 Bridge, "Land education and reconciliation."
34 University of British Columbia, Campus + Community Planning, *Musqueam Street Signs at UBC*, last modified January 17, 2020 https://planning.ubc.ca/musqueam-street-signs
35 Musqueam, 2020.

36 Sandra Styres, "Land as first teacher: A philosophical journeying," *Reflective Practices* 12: 721, 2011.
37 Bill Anthes, *Edgar Heap of Birds*, (Durham, NC: Duke University Press, 2015).
38 William Smith, "In the studio: Hock E Aye Vi Head of Birds," *Art in America*, last modified January 17, 2020. https://www.artnews.com/art-in-america/features/in-the-studio-hock-e-aye-vi-edgar-heap-of-birds-63298/.
39 Calderon, "Speaking back to manifest destinies," 28.
40 Megan McGinty and Megan Bang, "Narratives of dynamic lands: Science education, Indigenous knowledge and possible futures," *Cultural Studies of Science Education* 11, (2): 473, 2016.
41 Audra Simpson, *Mohawk Interruptus: Political Life Across the Borders of Settler States*. Durham, NC: Duke University Press, 33, 2014.
42 University of British Columbia, UBC Centennial, *The Hidden History*, accessed January 17, 2020. http://centennial.aboriginal.ubc.ca
43 Jordan Wilson, *A Walking Tour of Musqueam House Posts at the University of British Columbia,* last modified January 17, 2020. https://belkin.ubc.ca/wp-content/uploads/2019/05/FINAL_WEB_REVISED.pdf
44 Musqueam, 2020.
45 Jan Hare, "To know papers: Aboriginal perspectives on literacy."
46 Jan Hare, "They tell a story and there's meaning behind that story": Indigenous Knowledge and Young Indigenous Children's Literacy Learning, *Journal of Early Childhood Literacy* 12, (4): 392, 2012.
47 Zane Ma Rhea and Lynette Russell, "The invisible hand of pedagogy in Australian Indigenous studies and Indigenous education," *Australian Journal of Indigenous Education* 41 (1), 2012.
48 Calderon, "Speaking back to manifest destinies."
49 Stein, "'Truth before reconciliation,'" 164.
50 Ibid, 165.
51 Ibid, 168.
52 Ibid (168)

References

Alozzi, Raneem. "Ryerson unveils plaque revealing its racist history," *The Eyeopener*, June 28, 2018, accessed January 15, 2020. https://theeyeopener.com/2018/06/ryerson-unveils-plaque-recognizing-its-racist-history/.

Anthes, Bill. *Edgar Heap of Birds*. Durham: Duke University Press, 2015.

Barnhardt, Ray and Angayuqaq Oscar Kawagley. "Indigenous knowledge systems and Alaska native ways of knowing," *Anthropology & Education Quarterly* 36 (1): 8–23, 2005.

Basso, Keith. *Wisdom Sits in Places: Landscape and Language among the Western Apache*. Albuquerque: University of New Mexico Press, 7, 1996.

Bridge, Christine. "Land education and reconciliation: Exploring educators' practice" (PhD diss., University of British Columbia, 2018).

Burton-Christie, Douglas. "Place-making as contemplative practice," *Anglican Theological Review* 91 (3): 347, 2009.

Calderon, Dolores. "Speaking back to manifest destinies: A land education-based approach to critical curriculum inquiry," *Environmental Education Research* 20 (1): 24–36, 2014.

Daka, Favour. "Nothing is set in stone: Colonial statues on campus," *The McGill Tribune*, April 10, 2019, accessed January 15, 2020. http://www.mcgilltribune.com/opinion/nothing-is-set-in-stone-colonial-statues-on-campus-1004/.

Deloria, Vine Jr. and Daniel Wildcat, *Power and Place*. Golden: Fulcrum Publishing, 2001.

Ewert, Alan and Jim Sibthorp, *Outdoor Adventure Education: Foundations, Theory, and Research*. Champlaign: Human Kinetics, 2014.

Friedel, Tracy L. "Looking for learning in all the wrong places: Urban native youths' cultured response to Western-oriented place-based learning," *International Journal of Qualitative Studies in Education* 24 (5): 531–546, 2011.

Gaudry, Adam and Danielle Lorenz, "Indigenization as inclusion, reconciliation, and decolonization: Navigating the different visions for Indigenizing the Canadian Academy," *AlterNative: An International Journal of Indigenous Peoples* 14 (3): 218–227, 2018.

Hare, Jan. "To know papers: Aboriginal perspectives on literacy," in *Portraits of Literacy Across Families, Communities and Schools*, ed. Jim Anderson et al. Mahwah: Lawrence Erlbaum Associates, 243–263, 2005.

Hare, Jan. "'They tell a story and there's meaning behind that story': Indigenous Knowledge and Young Indigenous Children's Literacy Learning," *Journal of Early Childhood Literacy* 12, (4): 389–414, 2012.

Holmes, Cindy, Sarah Hunt, and Amy Piedalue, "Violence, Colonialism, and Space: Towards a Decolonizing Dialogue," *ACME: An International Journal for Critical Geographies* 14 (2): 540–570, 2015.

Ma Rhea, Zane and Lynette Russell, "The invisible hand of pedagogy in Australian Indigenous studies and Indigenous education," *Australian Journal of Indigenous Education* 41 (1): 18–25, 2012.

McGinty, Megan and Megan Bang, "Narratives of Dynamic Lands: Science Education, Indigenous Knowledge and Possible Futures," *Cultural Studies of Science Education* 11 (2): 471–475, 2016.

Metcalfe, Amy Scott. "What is difficult knowledge?," *The Difficult Knowledge Project* (blog), accessed January 12, 2020. http://blogs.ubc.ca/difficultknowledge/.

Minthorn, Robin Starr and Christine A. Nelson, "Colonized and racist Indigenous campus tour," *Journal of Critical Scholarship on Higher Education and Student Affairs* 4: 73–88, 2018.

Mullins, Philip, Gregory Lowan-Trudeau, and Karen Fox, "Healing the split head of outdoor recreation and outdoor education," in *Routledge International Handbook of Outdoor Studies*, ed. Barbara Humberstone*et al.*Abingdon: Routledge, 49–58, 2016.

"Musqueam," A Living Culture, accessed January 14, 2020. http://musqueam.zenutech.com/.

"Musqueam," Who We Are, accessed April 5, 2020. https://www.musqueam.bc.ca/our-story/who-we-are/.

Scully, Alexa. "Unsettling place-based education: Whiteness and land in Indigenous education in Canadian teacher education," *Canadian Journal of Native Education* 1 (38): 80–101, 2015.

Simpson, Audra. *Mohawk Interruptus: Political Life Across the Borders of Settler States*. Durham: Duke University Press, 33, 2014.

Smith, Linda Tuhiwai. "Foreword: Keeping a decolonising agenda to the forefront," in *Decolonisation in Aotearoa: Education, Research and Practice*, Jessica Hutchings and Jenny Lee-Morgan, eds. Wellington, NZ: NZCER Press, ix–x, 2016.

Smith, William. "In the studio: Hock E Aye Vi Head of Birds," *Art in America*, September 25, 2017, accessed January 17, 2020, https://www.artnews.com/art-in-america/features/in-the-studio-hock-e-aye-vi-edgar-heap-of-birds-63298/.

St. Germain, Brenda. "Behind the colonial wall: The chains that bind resistance" (Master's thesis, University of Victoria, 2014).

Stein, Sharon. "'Truth before reconciliation': The difficulties of transforming higher education in settler colonial contexts," *Higher Education Research & Development* 39 (1): 156–170, 2020.

Styres, Sandra. "Land as first teacher: A philosophical journeying," *Reflective Practices* 12: 717–731, 2011.

Styres, Sandra, Celia Haig-Brown and Melissa Blimkie, "Toward a pedagogy of land: The urban context," *Canadian Journal of Education* 36 (2): 188–221, 2013.

Truth and Reconciliation Commission of Canada, "Truth and reconciliation commission of Canada: Calls to action" 2015, accessed January 15,2020. http://trc.ca/assets/pdf/Calls_to_Action_English2.pdf.

Tuck, Eve, Marcia McKenzie and Kate McCoy, "Land education: Indigenous, post-colonial, and decolonizing perspectives on place and environmental education research," *Environmental Education Research* 20 (1): 1–23, 2014.

University of British Columbia, Campus + Community Planning, Musqueam Street Signs at UBC, accessed January 17, 2020. https://planning.ubc.ca/musqueam-street-signs.

University of British Columbia, UBC Centennial, *The Hidden History*, accessed January 17, 2020. http://centennial.aboriginal.ubc.ca.

Wilson, Jordan. "A walking tour of Musqueam house posts at the University of British Columbia", accessed January 17, 2020. https://belkin.ubc.ca/wp-content/uploads/2019/05/FINAL_WEB_REVISED.pdf.

8 The Haudenosaunee Thanksgiving Address and its relevance for futures and learnings

Roberta Hill

Introduction

In this pandemic period, how might we think about our future relationship to the Earth and sky so human beings learn to be actively beneficial to the Earth's sustaining life? In an Indigenous future vested in planetary recovery for humankind and the Earth, an understanding of The Haudenosaunee Thanksgiving Address may help people cope with the threats of modernity: pandemics, climate change, extinctions, nuclear power, resource depletion, war, famine and institutional crises. The practice of Thanksgiving is a powerful ritual that has helped the Haudenosaunee to endure hardships, loss and grief. Through this ritual, they remember their responsibilities to support the interrelatedness of life, including the Earth. I will use the form of the Address that is available worldwide, as will be explained in the chapter. My intention is to share what I believe is valuable and meaningful, given that this ritual is ancient. I am not a cultural spokesperson, but an Oneida who finds in this ritual an enriching meaning and healing perspective to share with others. When I taught my classes about the ritual, students sometimes thought I was expecting them to pray or appropriate the practice… that was not my intention and not my intention in this chapter. The Thanksgiving Address is a practical way to release people from their social tensions and have them consider their greater responsibility for life itself.

In this chapter, I focus on four major themes relating the Thanksgiving Address: time, place, the future and learning. These themes are Changing Our Minds; Spirit as Life; Time, Place, Memory; Our Indigenous Future. In Changing Our Minds, I'll discuss what the ritual is, its availability, and share how it creates collective energy to address problems, the greatest being death and grief. Spirit as Life shares what I've learned about the ritual from speaking with more knowledgeable Oneidas who speak the Oneida language. Speaking with Mr. Bob Brown and Chief Justice Hill gave me a deeper understanding and esteem for this ritual. In Time, Place, Memory, I'll focus on the Oneida Nation of Wisconsin, briefly mentioning the history then reframing "history" as related to time and place. In the final section, I'll explore the precarious nature of Our Future? Indigenous and Otherwise, and the need to accept our responsibilities for this fragile planet, our only home. The future now holds the invisible disease, COVID-19, caused by SARS Cov-2, a coronavirus that is changing global, national, tribal, and local economies as well as social interactions worldwide. Covid-19 may signal the possibility of new beginnings, if humankind changes its path by recognizing that war, fossil fuel extraction, capitalism and habitat destruction keep us locked into a recurring pandemic future.

Changing our minds

The Thanksgiving Address, also called "Words Spoken Before All Others," begins and ends Haudenosaunee ceremonies and meetings to discuss a problem. "Haudenosaunee" refers to the idea of a Longhouse with five fires, one for each Nation, stretching through the homelands of the Mohawk who are farthest east, the Oneida, the Onondaga being the central fire, the Cayuga, and Seneca on the far west. Prior to colonization, within palisaded villages were 30 to 150 homes that looked like Quonset huts, 15 to 20 feet tall, with two long outer walls covered with bark, a curved roof with smoke holes and movable openings for sunlight. At either end were entrances that could be closed with panels. Down the central corridor burned fires, one for each sister and her kin; built into the frame on either side were raised platforms for sleeping.[1] The Longhouse as a metaphor meant that all the nations under its roof were relatives. These five nations spoke related languages and the people were relatives through their clans. The nations themselves were also conceptualized as being older or younger brothers. After years of war, The Peacemaker, and his speaker, Hiawatha[2], united these nations and established the Great Law of Peace. When the Tuscarora War ended in 1713, the Tuscarora left their homeland in North Carolina and migrated north to become the sixth nation of the League.[3] The Haudenosaunee uphold an ancient tradition of reminding their people to be grateful for Mother Earth and all that creates life. The Thanksgiving Address is that reminder and an ancient and contemporary custom within the Longhouses and in the Native Nations.

The Thanksgiving Address as a ritual is considered to be as old as the Haudenosaunee migration to their homelands in New York and Canada. This migration occurred nine thousand years ago, the length of time that these Nations have been in their homelands. The Address is considered to be part of the original learnings to show people how to deal with grief and death by directing their attention from grief to the continuation of life.[4]

A readily available text that shares this ritual, *Greetings to the Natural World Ohen: ton Karihwatehkwen: Words Before All Else*, was inspired by *Tekaronianekon* (Jake Swamp) and collectively written by John Stokes, *Kanawahienton* (David Benedict), *Rokwaho* (Dan Thompson), with artwork by *Kahionhes* (John Fadden), published by the Tracking Project and Six Nations Indian Museum (1993). It's important to note how the text became published, because the ritual is spoken in one of the six native languages, and due to colonization and the suppression of our languages, having a speaker of the language in the community is a treasure. Having this version published also gives non-speakers, such as myself, a greater understanding of this tradition.

In the 1990s, *Tekaronianekon*, Jake Swamp, a spiritual leader of the Wolf Clan, envisioned a way to share the Thanksgiving Address with the world. He travelled worldwide, planting Trees of Peace and teaching about the Haudenosaunee Great Law of Peace and the Thanksgiving Address. According to Dave Fadden of the Six Nations Historical Museum in Onchiota, New York, Jake Swamp, *Kanawahienton* (David Benedict), *Rokwaho* (Dan Thompson), *Kahionhes* (John Fadden), and John Stokes met, reviewed various versions of the Thanksgiving Address, and discussed how to publish a version. John Stokes took the versions translated into English and wrote three sentences for each section of the address. From the English version, *Kanawahienton* and *Rokwaho* translated the sentences into Mohawk. John Fadden created the artwork; Andrew Main and Michael Motley designed the small book, first published in 1993.[5] *The Thanksgiving Address: Greetings to the Natural World* is available from the Tracking Project in ten languages: English, German, Swedish, Japanese, Portuguese, Spanish, Bisayan, French, Hawaiian and Italian, all with the Mohawk version included. According to the Tracking Project web site, over the last seventeen years, 100,000 people have bought copies.[6] It is the source most often found on the Internet. When I share my interpretation, I will be using this text.

According to John Stokes, Jake Swamp wanted a version that would reach people worldwide. The group of Mohawk leaders, Tom Porter, Oren Lyons, Ed Benedict and Jake Swamp, wanted John Stokes, who is non-Native, to write a version accessible to English speakers. Dan Thompson, who ran the Mohawk Language Program, feared Mohawk would be a dead language in 20 years. The group met and went through the versions in order to determine what they wanted to use, at times altering the English phrases to match the Mohawk, at other times determining what to exclude. Jake Swamp wanted the Thanksgiving Address translated in many languages, so that young people in the future may understand it, each in their language. With the help of these leaders' efforts, both the Mohawk language and the Hawaiian language are growing in the number of speakers.[7]

The Oneida Nation Cultural Heritage Department also sponsored a video of the Oneida version along with images of the natural world. Mr. Bob Brown, a native speaker and the Cultural Adviser for Oneida Language Revitalization Program, produced the video, which is no longer available.

In his book, *Kayanerenko:wa: The Great Law of Peace, Kayanesenh* Paul Williams opens his book with "Ohe'n: ton Karihwatekwen: Words Before All Else" and explains that "This gratitude reminds us, too, that we humans are no more important than the other living parts of the world."[8] This would be one source for those readers interested in a deeper understanding.

Chief Judge Hill of the Oneida Nation of Wisconsin explained that The Thanksgiving Address opens our rituals. It expresses the recognition of our place on the earth and all the beings that support and sustain life. It is intended to focus the community's attention to the interconnected relationships between the Earth, sky and all that is in-between, to generate a communal vision of these relationships, and to have people let go of their resentments and hostility in order to generate gratitude for the Earth's power, so they will act with positive communal effort to gather together and send forth the energies for sustaining life.[9]

The Thanksgiving Address reminds the people that whatever their hostilities, jealousies and tensions, they need to pay attention to the moment in time and the place where they are gathered. They need to recognize their individual energy contributes to a greater purpose, using a phrase that is repeated: "Now our minds are one." In talking with Chief Judge Hill, we discussed how the cultural values expressed differ from those of people who ask for something from God or the Creative Forces. Unlike the worldview that focuses on control and appeasement, on getting the Creator/God to provide help in obtaining a heart's desire, the Thanksgiving Address brings people together to remind them to observe peace as a priority, that the natural world creates responsibilities for many life forms, from the Earth as mother to all of the life forms, including sun, moon and stars. Peace makes life continue. By listening to the speaker, the listeners clean up their attitudes and clear their minds so they can collectively focus on the problems facing them.[10]

Haudenosaunee communities have endured centuries of trauma. The Thanksgiving Address gives them the way to move beyond helplessness and despair. Helplessness and despair lead to death. The Address reminds them to pay attention to what they have and to express gratitude to one another and to the energies that sustain life. This is appreciation, not gratification. So, by the time the speaker has reminded the people of what is before them, how the natural world has order and beauty they appreciate – the wind through the forest churning the layers of leaves, the warmth of sunlight, the laughter of children, all the forms of life, inanimate and animate, become present to the people gathered together. With that worldview, the people understand their responsibility is to continue life in all its complex and varied manifestations.

Death and grief are also implicit in the Thanksgiving, because death is accepted as a natural outcome of life, each of us given only so much time. Grief can be a disruptive social process, especially on a large scale, so the Thanksgiving Address alleviates the dread of death, layered by physical pain, by the emotional pain of leaving loved ones, and on a deeper level, by the pain of leaving the Earth. When facing death, the need for the Earth, for soil, rocks, trees, birds, may be a hunger more confounding than the need for air, water, food, and sexual intimacy. What is needed, which the Address provides, is the communal faith that the Earth supports humankind and loves us back. In my view, when death comes on a massive scale, as in pandemics, warfare and the theft of land, in the hardship of forced migrations, in brutality and suffering, people may fight back, migrate, or grow despondent. A grieving people's response to collective death and trauma can devolve into violence and chaos.

The Thanksgiving Address brings people into another place and time, to help them remember that life continues as their minds are cleansed with their gratitude, and their collective responsibility to see that life continues.

Spirit as life

When I asked Bob Brown, an Oneida Native Speaker, how to understand this ritual, he explained that the Thanksgiving Address goes back to the Oneida Creation story. The story shares the belief of each person having spirit and life. When a group of people come together and focus their minds, their spirits/lives grow, as does the energy of those beings they address. Bob Brown condensed the Creation Story to show the connection between the two. Sky Woman was a celestial being with powers human beings don't have. She was pregnant and after falling to the watery sea, she gave birth to her daughter, who grew rapidly and matured. Her daughter went out, slept and when she woke, she found two arrows on her belly: one straight and well-made, the other arrow crooked with an unevenly chipped arrowhead. Her mother knew the arrows meant her daughter was pregnant. The twins she carried argued about being born. One came out as all babies do, by going down. The other went toward the light in his mother's mouth, but couldn't go that way, and came out under her armpit, killing her. Sky Woman buried her daughter in the Earth, and from her body came all the life around us. She is Mother Earth. As celestial beings the twin boys grew quickly into men. One created human beings from different substances: foam on the sea became white people, sand on the shore became yellow people. He went into the forest and with rich black soil created black people, and from brown clay created red people. His twin created monkeys and apes from tree branches. All beings have the spirit of these origins from the celestial realm whose path is the Milky Way. Without distinguishing between "animate/inanimate" all life on Mother Earth and above, including Sky Woman (the Moon), and Elder Brother Sun, have the same spirit. All that is on Earth and in the cosmos keeps its relationship to the place beyond the Milky Way. If we express gratitude and recognize our relationship to life on Earth, if we give attention to the meaning behind these relationships, we feel harmony. Bob Brown translated the word for "spirit" as "*atunhetsla*" which also means "life". Life and spirit are one and the same; in this cosmology, the soil and the tree come from the same celestial source – a star or group of stars beyond the Milky Way. In the video, I noticed that the order of addressing the plants and elements differed from the order expressed in the book. He said that each speaker decides on the order, because the tradition is fluid and considerate of the occasion and circumstances.[11]

I spoke with Chief Judge Gerald Hill, Chief Appellate Judge for the Oneida Nation of Wisconsin. He speaks Oneida and explained the address may be long or short, can go on for a long time if the speaker is grateful for all he or she wants to name, all the medicines, all the trees or animals, as a progressive pattern that calls for gratitude more than asking for something. Those listening to the speaker become part of an intricate process and keep open minds and hearts, so they more readily follow what the discussion of the problem or ceremony involves. Chief Judge Hill explained that to listen to an accomplished speaker share the intricacies of these relationships in the Oneida language increases the listeners' awareness of the need for patience, care, and reason. The speaker may also make the address short. The listeners experience the creative use of language to establish harmony and open-mindedness; they absorb the music of the words and the philosophy of gratitude. The Address gives listeners an ideal, showing the flow of relationships that sustain life, including human relationships. Chief Judge Hill explained that the Address is intended to bring out the best in people, giving them a way to pay attention and have an open mind, to acknowledge and thank each other and the natural phenomena that sustain life.[12]

The Thanksgiving Address incorporates the future by increasing our awareness of the balance within our community, and the need for differences. Not all grasses or birds are the same, so it is with human beings. To learn more about how this feels, I will use the Thanksgiving Address published by the Tracking Project in order to interpret what I notice about its function, its beauty and the depth of its power. This is only one version. The Thanksgiving Address begins by having the people be grateful for each other and to bring their minds into harmony. In this version, the speaker begins with the earth and ends with the Creator, bringing forward into the listeners' ears the intricacies of life. The life forms being addressed are "The Earth Mother," "The Waters," "The Fish," "The Plants," "The Food Plants," "The Medicine Herbs," "The Animals," "The Trees," "The Birds," "The Four Winds," "The Thunderers,[13]" "The Sun," "Grandmother Moon," "The Stars," "The Enlightened Teachers," "The Creator," and "Closing Words".[14] Each life form has a responsibility, for example, water strengthens us, quenches thirst, and comes in many forms. The listeners may recall an image from their own experience of what is being addressed, for example: "We turn our minds to all the Fish life in the water. They were instructed to cleanse and purify the water. They also give themselves to us as food. We are grateful that we can still find pure water."[15] For Oneidas who had a great resource in Oneida Lake and the many rivers, recognition of the fish, which was a main food source in the early Spring, reminded the people of the need for pure water.

The relationships among people, life forms and the land were based on kinship. Tom Porter, a Mohawk activist who established *Kanatsioharake*, a Mohawk community west of Albany, N. Y., explained to me that in Mohawk, the words for "sun," "moon," and "earth" state kinship relations: "Elder Brother Sun," "Grandmother Moon," "Mother Earth," so in the Mohawk language, a speaker could not separate himself/herself from a close relationship to life. Another example from the text addresses the sun: "We now send greetings and thanks to our eldest brother, the Sun. Each day without fail he travels the sky from east to west, bringing the light of a new day. He is the source of all fires of life."[16] In this passage, the people see the sun as a living being, a brother who cares for them and all life on Earth. Light moves everywhere at once, reaches even the depths of the sea in its movement of currents, so the sun's power provides for humankind and the Earth Mother as well.

In this version, readers/listeners may also think of the layers of intricate responsibilities, the "ecology" of life enclosing and sustaining them. It calls for the people's collective attention and gratitude, reminding the listeners of how all life communities need to be acknowledged, how every layer of the place where the people stand presently is intimately related to other layers, from the Earth to the cosmos. It brings forth a different imaginary of the future, changing the inanimate conceptions of settler-colonial society's materialism to conceptions of the kinship real human beings must have with all that is around them in their homeland, Earth to sky.

What heartbreak to lose quetzals and paladins, hummingbirds and finches, frogs and sea turtles, porpoises and whales, racoons and bats, and the millions of species that will be extinct due to fossil fuel extraction and climate change. If we were to express our gratitude for them in a collective setting, becoming aware of how each one takes on an ongoing and present responsibility within the flow of life, we might not allow such losses, since a loss in one community of life forms challenges the responsibilities of all other forms, because the interrelationships among the plants to trees to animals become clear to an open mind. I can't address the beauty of the ritual as it is spoken in Oneida. In the printed version, a learner/reader may find a different perspective, for it mentions animal life as teachers; trees as having families; birds as teaching us to enjoy life; the Thunderers, as sky forces and grandfathers talking together when they replenish the waters. When I read this small book, I have to stop to listen and feel the birdsong, the wind, feel the sunlight on my arms. My mind is changed from reading through the book and when attending the ritual, to listen to the Oneida language. The listeners hear the phrase of bringing their minds together to greet and thank The Elder Brother Sun, Grandmother Moon, and the Stars. In their awareness, the listeners have travelled from the ground they stand on to the sky above, or whatever order the speaker creates. They listen to the speaker remind them to thank "The Enlightened Teachers" who remind the people "of the way we were instructed to live as people".[17] The beauty of these relationships throughout is one of learning and being taught about real human beings' responsibility to the Earth and all she provides. Whether one is five years old or 85, being reminded of how we are to live and what we need to keep our minds open, to feel the beauty and power of each life form as a collective and as an individual may inspire listeners to feel the wonder and beauty of the Earth and life itself.

The Thanksgiving Address strengthens the power of memory, for not only must the speaker remember, but the people also need to recall their own experience when the speaker reminds them of their relationships to other life forms. Memory was and still is a phenomenal resource. Oneida elders told me how their grandparents reminded them to remember what was said. The ritual focuses on life as a continual spiritual force, one which we must remember, especially in times when death and hardship come. The ritual brings the energies of all the people together, relieving isolation and loneliness, and providing the energy and power to overcome fear. The dread of death was not an unfamiliar experience to those who established this ritual. They must have known as warriors how grief can destroy kin and community. The power of love and the awareness of responsible action strengthens and helps people to stay resolute in their gratitude for *atunhetsla*, spirit meaning life.

Time, place, memory

When British colonists met the Haudenausaunee, they learned about the ancient story of the Great Law of Peace and observed the ancient practice of Thanksgiving. *The Basic Call to Consciousness, the Haudenosaunee Address to the Western World* expresses the ethnical framework of The Great Law of Peace and the beliefs and assumptions for living as real human beings:

> Ours is a Way of Life. We believe that all living things are spiritual beings. Spirits can be expressed as energy forms manifested in matter. A blade of grass is an energy form manifested in matter — grass matter. The spirit of the grass is that unseen force which produces the species of grass, and it is manifest to us in the form of real grass.
>
> All things of the world are real, material things. The Creation is a true, material phenomenon, and the Creation manifests itself to us through reality. The spiritual universe, then, is manifest to Man as the Creation, the Creation which supports life. We believe that man is real, a part of the Creation, and that his duty is to support Life in conjunction with the other beings. That is why we call ourselves *Ongwhehonwhe*— Real People.
>
> The original instructions direct that we who walk about on the Earth are to express a great respect, and affection, and a gratitude toward all the spirits which create and support Life. We give a greeting and thanksgiving to the many supporters of our own lives — the corn, beans, squash, the winds, the sun. When people cease to respect and express gratitude for these many things, then all life will be destroyed, and human life on this planet will come to an end.[18]

The Haudenosaunee position paper was presented at a conference at the United Nations in 1977 in Geneva, Switzerland. The address relates how the Six Nations acknowledge "the people" meaning all life that sustains humankind, in order to remember and preserve the past, present and future unfolding of spiritual energies. The end isn't ordained. It is a well-reasoned outcome of failure to acknowledge and respect the sources of life. "Reason" isn't only the prerogative of European philosophers; it may be studied in the Enlightenment, but reason belongs to humankind. In my view, in meeting with Native American confederacies over several hundred years, European colonizers were influenced by leaders who stressed the power of thought. In this brief section, I want to raise the question of the power of colonization to place European and American ideas as the guides to time, place, future and learning.

I grew up with the image of Europe and the U.S. culture inheriting "Western Civilization" as the primary force for how things are. In this view, the Greeks were the foundation of modernity. Eurocentrism diminishes our understanding of the central role that Native Nations had in shaping both democracy and the world as it is now. One outcome for learning is for Indigenous youth and youth worldwide to learn the complexity of their own identities and challenge those assigned to them, particularly racial identities. In his in-depth history of racial formation, Howard Winant explains that "only by assigning racial identities to all beings, only by generalizing a racial culture globally, was the new world order [modernity] able to constitute itself as a social structure at all."[19] Discovering one's familial and cultural history helps learners to become more critical of others ascribing stereotypical characteristics to them. By learning about their own particular identity, students can also accept the unique identities of those they encounter.

The immense natural riches of the Americas, Africa and Asia financed European imperialism. Alexander Koch and colleagues at University College London researched the global effects of climate due to the genocide of 56 million out of a population of 60.5 million in the Americas. Their research explains Europe's little ice age, 1500–1800, was due to a two degree drop in global temperature caused by the loss of agricultural production in the Americas. Koch and his colleagues claim the ice age's decline began by 1800 due to industrialization.[20] Professor Bruce Fleury explains that between 950–1250, Europe's Medieval Warm Period, agriculture and population growth created flourishing human communities. However, in the subsequent 500 years, 1350–1850, Europe experienced the Little Ice Age. He includes 1350 to 1400 because in that period the black rat brought plague. From the 14th and 15th century on, agriculture collapsed, bringing famine and malnutrition. Millions died, because plague, the Black Death (three different forms), increasingly infected European populations. The vector was bacterium living in a flea's gut. Thus, the fleas in black rats carried plague. Rivers of rats flowed into and out of the starving villages and cities filled with garbage and sewage. Dr Fleury explains that "Ring around the rosie ... ashes, ashes, we all fall down" was from the plague of London.[21] The Pied Piper of Hamilton isn't really a children's story; it refers to an historical experience. Rather than being exceptional, European living conditions drove colonizers out of Europe and led to the invasion and colonization of the immensely diverse Native Nations throughout the Western Hemisphere, which we call "Turtle Island". The Western Hemisphere is the continent that extends from the Artic to Tierra del Fuego. The hemisphere holds an incredible cultural, linguistic and political diversity of nations, millions strong and brilliant, nations which changed and were challenged during historical encounters with European colonization, with the names the people called themselves changed, with their ways of life studied, appropriated or held secretly in their families, with their lands taken by force and by treaties, with forced migrations and removals, with the recovery of traditions and languages in flux, and with their own resolute voices stating again and again the intention to survive and remember. It is hard to teach what the majority of Americans don't believe and deny. It is hard to learn that your invisibility as a surviving Indigenous person is due to genocide. To gain understanding, we must put European colonization and imperialism in perspective. Thus, I applaud the way Alexander Koch and his colleagues' current research challenges Eurocentrism.[22]

The wealth from Indigenous resources and forced mass labour, both slavery and peonage, amalgamated Europe's nation-states with us/them ideology.[23] Feudal warlords established hierarchical social orders, so that one member of proper blood-lineage held absolute power over peoples, territories and colonies. The monarch roused people to overcome foes and glory in looting, to obtain land and wealth, subjugate, slaughter, infect or drive off captured populations, then search for more enemies. This colonial mindset needs to change and is being challenged. In contrast, concepts of democratic government grew from colonial contact with Haudenausaunee and Eastern woodland confederacies who also went to war, but whose concepts of government and rule were egalitarian and collective, based on the assumption that "all people stood the same height" since the ability to reason and be aware makes collective action valuable, ethical and necessary for sustaining life.

Dutch, British and French colonization led to migrations and forced removals from traditional territories. Catholic and Protestant missionaries fractionated traditional communities. During the Revolutionary War, the Oneida helped George Washington's army by providing him corn when he was planning on crossing the Delaware River and turn the tide of war toward the Revolutionaries. In turn, Washington attacked Oneida and other Haudenosaunee towns, earning him the name, "Town Destroyer". The loss of our homelands and the ways in which the Oneida Nation of Wisconsin pursued justice for over two hundred years is one of many complex stories in Oneida history. The Oneida now reside in four areas: the Oneida Nation in the homeland of New York State; the Oneida Nation of Wisconsin; the Oneida at Six Nations of the Grand River, Ontario; and the Oneida of the Thames in Southwold, Ontario. Other Haudenausaunee Nations live in communities in New York, Oklahoma and Canada.

In a fraught history of hope and betrayal, the Oneida bought land from the Menominee and settled near Green Bay, Wisconsin in 1822. Various parties migrated and settled west of the old French port, and their land was diminished by treaties, and government policies throughout the 19th and early 20th centuries. The Oneida Nation of Wisconsin has worked to resolve the intense and imposed poverty and lack of community institutions brought on by federal policies. Due to the Native resistance of the 1960s to the present, the Oneida Nation is now one of the major employers in the Green Bay area and has created positions in many divisions of tribal enterprises for many tribal members to be employed. Due to the federal limits on reservation boundaries like many

Native Nations in the U.S., The Oneida still contend with non-Indian neighbours who had historically profited from obtaining Native land and resources. Unlike a township or city, the reservation boundaries contain a small number of people out of the entire tribal population. The land itself is a checkerboard of Native and non-Native ownership. Those neighbours have tried in multiple ways to deny the Oneida Nation of Wisconsin's political status as a federally recognized nation with a right to self-determination. The Oneida of Wisconsin have much to offer these neighbours, if they can open their minds toward understanding.

Our future? Indigenous and otherwise

For five centuries in an evolving dynamic process, mining for mineral wealth, fossil fuel extraction, colonization and capitalism formed an interlocking system where greed obtains value and power. This system continues, bringing death to animals, like the sea cucumbers, bats, pangolins, panthers, to name a few. Dutch elms, ash trees, oaks and other trees also have been wiped out. Elizabeth Kolbert in her Pulitzer Prize winning book, *The Sixth Extinction: An Unnatural History*, traces the long, painful history of this destruction, which she links to earlier extinctions and the history of geology. The current extinction grew from the colonizers' practice of capture, collection and categorization, driven by profit.[24] In his Earth Day Session, available on YouTube, Dr Jonathan Patz explains the "drivers" of the COVID-19 pandemic: habitat destruction leading to animal vectors mingling their viruses to create new, natural mutations; climate change which alters heat and moisture, so that deadly vectors like mosquitos increase. In other words, the current civilization causes pandemics. He argues for a "planetary health framework" for addressing the likelihood of ongoing pandemics. He explains what drives pandemic outbreaks statistically: habitat destruction brings animal species into contact, enabling pathogens to "spill over". The destruction of ecosystems worldwide may bring all sorts of pathogens into play. The Planetary Health framework intends to stop the disruption of natural systems.[25] It's one path towards change. Deeper change is needed, a change of consciousness. Indigenous peoples reasoned with the invaders, challenged and fought against colonization, protested and altered their political relationships with governments and international institutions and continue to challenge the notion that the Earth belongs to those with wealth and power. How are the elite going to accept a Planetary Health framework if it fails to stop the accumulation of capital?

My original introduction for this chapter, written in September 2019, concerning the Future and Teaching, became irrelevant when the current future altered the imaginary ones brought to worldwide audiences in films like Aquaman (2019) (EARNING 1 Trillion.)[26]; Captain Marvel (2019) (1 Trillion.)[27], and Avengers: Endgame (2019) (2.7 Trillion.)[28]. In these films, a few heroes with extraordinary powers stop social collapse with techno warfare. "Getting back to normal" was achieved through violent means with sources of sustenance magically continuing, a colonizer's contradiction. These movies grossed trillions, their conflicts foregrounding the problem of power and rule by assuring their audiences that one or several young folks could disarm one or more evil enemies. The films engage their audiences with the struggle between absolutism and democracy, but the community is a backdrop, not a force for change. COVID-19 challenges the single hero assumption. The phrase "We're all in this together" has been a rallying cry, but has yet to unfold as a substantial, system-altering course of action.

Virologists are still studying the current virus. We must stay home, use masks and stay six feet from all but family. The invisible strip of life has stopped traffic, cleared the air in highly populated cities, highlighted class and racial oppressions within global and U.S. socioeconomic structures, and in the U.S., it has caused more people to be out of work than in the Great Depression. It has made us isolated and dependent on technology for work, supplies and relationships. COVID-19 may signal the possibility of new beginnings, if we can recognize how war, fossil fuel extraction, capitalism and habitat destruction keep us locked into a recurring pandemic future. If teaching/learning is only available online, inequality in educational resources will exacerbate class and racial tensions. This is seen in the concern for "essential" workers: clerks, health care workers, delivery workers, migrant laborers, and all who must go out among the public to earn a paycheck. It has made daycare for working parents an essential service. Will the essential workers receive health care in the U.S.? Who will be vaccinated has not been addressed in a public forum. If only the elite get the vaccine, they will still be at risk, because if the poor die, the systems will still fail. Capitalism relies on the poor as much as the rich. Karl Marx argued that capitalism requires increasing numbers of poor people to act as the reserve labour pool and to scare all other workers, including the middle class, into meagre paychecks. The very wealthy work toward leaving Earth in a rocket to colonize Mars—a continuation of the false premise of colonization—that historic mercantile interests and subsequently corporate interests that industrialized the world can take from the Earth without regard for the consequences to all life.

In Indigenous futures, we might consider the people following the ethos of the Thanksgiving Address; if not using it directly, then as a means to discover in their own traditions and practices those rituals that give meaning to our lives on Earth, and that help us change the global system to make the Earth a sustainable home.

Conclusion

To create Indigenous futures means we take responsibility for life continuing, for feeling gratitude and generosity; for healing all that has been destroyed means we are healing ourselves, that there is no separation between humankind and the Earth that nurtures us. To share the Thanksgiving Address and increase an understanding of its importance was the vision of Jake Swamp, John Stokes and the Mohawk and Onondaga leaders Jake Swamp gathered. This vision has borne fruit in the continuing use of the small book that has been translated into languages that span the globe.

What have I learned from writing this chapter and what I have wanted to teach my readers? The Thanksgiving Address changes our minds from fear of death and loss to a vision of gratitude for life, for recognizing our responsibilities to make sure that life continues and to value all those life energies, the Earth, water, plants, animals, trees, birds, powers in the sky, the teachers who've shared their experience and wisdom, and the Creator whose spirit is with us as well. I learned about how long this Address has been among the Haudenosaunee and what the background is for its availability in the small book. In relating the history of colonization and of the Oneida experience, I foregrounded all of Turtle Island as a site of future teaching and learning. Finally, I addressed the present experience of a pandemic that has challenged us to recognize the need for changing how the world we've inherited works. We are all responsible for making the change to save the creatures who teach us, and the Earth that heals us. And I've learned the idea that life and spirit are one, in the Oneida language, *atunhetsla*, which I hope will continue to inspire Indigenous people to work toward a sustainable future on this nurturing Earth.

Notes

1 Daniel K. Richter, *The Ordeal of the Longhouse: The Peoples of the Iroquois Leave in the Era of European Colonization*. Chapel Hill: University of North Carolina, 18–1, 1992. Richter gives a more detailed account along with a picture.
2 Hiawatha, or Ayonwatha, translated as "He who combs" is not the figure in Longfellow's poem. He was the speaker for the Peacemaker and helped turn the tyrant, Tadadaho, into a sachem whose name and position continues in the Grand Council of the League.
3 Daniel K. Richter, *The Ordeal of the Longhouse: The Peoples of the Iroquois League in the Era of European Colonization*. Chapel Hill: University of North Carolina, 1992, 238–239.
4 John Stokes, Telephone conversation, June 8, 2020. John Stokes explained in his meeting with the Mohawk and Onondaga leaders, they established through archeoastronomy that the Peacemaker came to the Seneca during a solar eclipse one thousand years ago. The Peacemaker recited the Address which had been an ancient practice. For a more detailed account of early habitation, see Richter, 13–19 and 31.
5 Dave Fadden, Telephone conversation, August 28, 2019.
6 The Tracking Project, accessed September 15, 2019. https://www.thetrackingproject.org/
7 John Stokes, Telephone interview, June 8, 2020.
8 *Kayanesenh* Paul Williams, *Kayanerenko:wa: The Great Law of Peace*. Winnipeg, Manitoba: The University of Manitoba Press, xi, 2018.
9 Interview with Chief Judge Gerald L. Hill, September 9, 2019.
10 Chief Judge Gerald L. Hill, Telephone conversation, June 5, 2020.
11 Bob Brown, Interview, September 9, 2019. The Oneida Creation Story is available on the Oneida Nation of Wisconsin web site: https://oneida-nsn.gov/our-ways/our-story/creation-story/. See also: Demus Elm and Harvey Antone, *The Oneida Creation Story*. Lincoln, NB: University of Nebraska Press, 2000.

12 Interview with Chief Judge Gerald L. Hill, September 9, 2019.
13 "The Thunderers" refers to storm clouds that bring thunder, lightning and rain; they are grandfathers with thunder voices.
14 John Stokes, *Thanksgiving Address: Words Before All Else*. Correles, NM, no page number, 1993. See also: https://www.thetrackingproject.org/ accessed May 22, 2020.
15 Stokes, *Thanksgiving Address: Words Before All Else*, no page number.
16 John Stokes, *Thanksgiving Address: Words Before All Else*, no page number.
17 John Stokes, *Thanksgiving Address: Words Before All Else*, no page number.
18 Akwesasne Notes, *A Basic Call to Consciousness The Hau de no sau nee Address to the Western World*. Geneva, Switzerland, 1977. Accessed on the internet, Feb 17, 2020 https://ratical.org/many_worlds/6Nations/6nations1.html#part1a
19 Howard Winant, *The World is a Ghetto: Race and Democracy Since World War II*. New York: Basic Books, 30, 2001.
20 Alexander Koch et al., "Earth system impacts of the European arrival and Great Dying in the Americas after 1492" *Quaternary Science Reviews* 207: 13–16, 2019. Accessed May 15, 2020. https://globalnews.ca/news/4924534/little-ice-age-death-55-million-indigenous-people-colonization-study/ and https://www.sciencedirect.com/science/article/pii/S0277379118307261
21 Bruce E. Fleury, *Mysteries of the Microscopic World Course Guidebook*. Chantilly: The Teaching Company, 11-12, 2011.
22 A good source for understanding Eurocentrism is *J. M. Blout, A Colonizer's Model of the World: Geographic Diffusionism and Eurocentric History*. New York: The Gilford Press, 1993. For a text about Indigenous cultural diversity, see Gerald McMaster and Clifford E. Trafzer, eds., *Native Universe: Voices of Indian America*. Washington, DC: National Geographic, 2008.
23 Howard Winant, *The World Is A Ghetto: Race and Democracy Since World War II*. New York: Basic Books, 24-25, 2001.
24 Elizabeth Kolbert, *The Sixth Extinction*. New York: Henry Holt and Company, 2014.
25 https://www.youtube.com/watch?v=PWSiZIDHY9M
26 The Numbers, accessed 6/12/20 https://www.the-numbers.com/box-office-records/domestic/all-movies/cumulative/all-time
27 The Numbers, accessed 6/12/20 https://www.the-numbers.com/box-office-records/domestic/all-movies/cumulative/all-time
28 The Numbers, accessed 6/12/20 https://www.the-numbers.com/box-office-records/domestic/all-movies/cumulative/all-time

References

Akwesasne Notes. *A Basic Call to Consciousness The Hau de no sau nee Address to the Western World*. Geneva, Switzerland, 1977. Accessed, Feb 17, 2020https://ratical.org/many_worlds/6Nations/6nations1.html#part1a

Blout, J. M. *A Colonizer's Model of the World: Geographic Diffusionism and Eurocentric History*. New York: The Gilford Press, 1993.

Brown, Bob. Interview. Oneida Nation, Wisconsin. September 9, 2019.

Elm, Demus and Harvey Antone. *The Oneida Creation Story*. Lincoln: University of Nebraska Press, 2000.

Fadden, Dave. Telephone Interview. August 28, 2019.

Fleury, Bruce E. *Mysteries of the Microscopic World Course Guidebook*. Chantilly: The Teaching Company, 2011.

Hill, Gerald L. Interview Oneida Nation, Wisconsin. September 9, 2019.

Koch, Alexander*et al.* "Earth system impacts of the European arrival and Great Dying in the Americas after 1492," Quaternary Science Reviews207: 13–16, 2019. Accessed May 15, 2020. https://globalnews.ca/news/4924534/little-ice-age-death-55-million-indigenous-people-colonization-study/ and https://www.sciencedirect.com/science/article/pii/S0277379118307261

Kolbert, Elizabeth. *The Sixth Extinction*. New York: Henry Holt and Company, 2014.

McMaster, Gerald and Clifford E. Trafzer. Eds. *Native Universe: Voices of Indian America*. Washington: National Geographic, 2008.

Richter, Daniel K. *The Ordeal of the Longhouse: The Peoples of the Iroquois Leave in the Eroa of European Colonization*. Chapel Hill, NC: University of North Carolina, 1992.

Stokes, John. Telephone conversation, June 8, 2020.

The Numbers, accessed June 12, 2020https://www.the-numbers.com/box-office-records/domestic/all-movies/cumulative/all-time

The Tracking Project. Accessed September 15, 2019. https://www.thetrackingproject.org/

Williams, Kayanesenh Paul. *Kayanerenko:wa: The Great Law of Peace*. Winnipeg: The University of Manitoba Press, 2018.

Winant, Howard. *The World is a Ghetto: Race and Democracy Since World War II*. New York: Basic Books, 2001.

9 Brown, Red, and Black to the Futures

Ligia (Licho) López López and Gioconda Coello

We have started and abandoned this conclusion multiple times. Perhaps that captures the mood the spirit of writing is in when across the planet too many of us (Brown, Red, Black, and everyone else) are perishing in the COVID-19 pandemic. The majority of the work in putting the book together has taken place between the raging Australian forest and bush fires from early January 2020 and through to the pandemic in the first half of 2020. However, the foundational work on the book began well before the fires and the start of the pandemic because the conditions of planetary threat have been *taking place* much earlier than late 2019 (or whenever the virus is speculated to have taken hold of human bodies in a wet market or anywhere else on the face of the earth). The foundational work on the book began well before the conditions of planetary threat, before the world became the map, at least 65,000 years ago since our ancestors started worlding life otherwise.

For many, the virus has been a catalyst to *notice* the historically extensive and expanding colonial conditions whose everyday performance seemed to have been eluding the attention of many people for centuries. Often ableist metaphors are employed to describe governments, the privileged, etc. "being blind" to racism, or "being deaf" to the cry of the impoverished, which they can now recognize due to COVID-19. *Noticing*, however, is a political act beyond ability. It is an act that debilitates the colonial complex which supplies the presumed selected optimal model of human[1] with health, safety, and protection which translates into time, life, and *their* future. *Not noticing* is convenient and profitable. The many revelations that have been mushrooming in OpEds, short and long essays, and which are slowly appearing in the academic publishing world in journals and books, are nothing but the future as we (the "deselected" kind, *les damné de la terre*)[2] have already known it.

What COVID-19 has brought is a compressed reproduction of long histories of disregard for Brown, Red, and Black lives. Indigenous lives once limited to reservations continue to be threatened now with the highest rates of COVID-19 infections in the United States. The Colombian Amazon continues to face genocide with the highest rates of infection in the nation as the rest of the country suffers from the ruthless massacre of Indigenous and Afrodiasporic community leaders. In Ecuador, Guayaquil streets are populated with the dead bodies of the hard-working class. As we write this, hunger is taking the people of Chile—particularly those impoverished, the Mapuches, and the made up "other"/migrants—to the streets in protest. But let us pause and *breathe* here. *Breathe, as George Floyd (RIP) teaches us, is a cry for futurity, the one Black people have built for centuries in the struggle for freedom to exist in a profoundly antiblack world.* Before we are tempted to go around the world sampling the spread-out devastation that COVID-19 has not precisely created but is indeed highlighting, let us return to the organizing intentions of this collection of chapters: Indigenous time and the futures.

To illustrate some of the points we are making here and to bring the discussion closer to education and schooling more specifically, let us go further back in time. What follows is a fictional story, written by Ligia, in two parts. By the Sinú river where I grew up (Ligia speaking), story was my education, my initiation into reading and making worlds that went and go beyond the limits and limitations of school, any school. On my grandmother's lap, on the rocking chair, outside *el techo de palma*, and under the starry skies of warm Caribbean nights, I had many questions and my grandmother had stories. We all had Gabo (Gabriel Garcia Márquez) and many more unpublished storytellers with magical realism and unnamed literary genres that stirred our imagination, and that was a revolutionary act of future(s) making taking place.

★★★★

Ancestral 2020 vision: Chronicle of a death foretold[3]

Part I

Wati't, tell me that origin story. The one where another life begins in the future.[4]

The year is…

never mind. We do not count time that way. Not in years… well at least not in Gregorian years. That is something that will begin with beings who will call themselves **Humans**.[5] They will live in the future. Tens of thousands of years from now, in the year 2020, there will be death unlike no other.[6]

Humans will be peculiar creatures, innovative beings who will conceive of life like no other being. They will invent abstract but consequential concepts such as the **self, others**, and **the individual**. In calling themselves Human, they will separate themselves from "other" beings.[7] They will invent **Difference**[8] to fragment themselves even further. Colour coding Human bodies will be fashionable for hundreds of years before 2020.[9] They will call us **Indios, Natives, Aborigines, Black, the n word**, and many, many more names.[10] Oh they will get real creative![11] They will invent sciences to carry out the business of coding, so anatomy will be used to binarize bodies based on extra or missing body mass, and psychiatry will act on the mind, and the body, to invent disabilities. They will create classes and a senseless system called capitalism[12] through which they will trade Human flesh and bones.[13] They will do this using the colour and other codes their many sciences will advance for decades. They will invent a deity called **The Economy** which millions of them will worship. Unknowingly, or perhaps not, their fearful devotion to The Economy will bring them devastation and perennial death.

One of the most remarkable achievements of their ingenuity will be called **Schools**. They are an amalgam of all the things that the Humans will invent. As architects, their sciences will make important contributions to the school edifice. The colour codes, the binaries, the disabilities will all be carried into the building. With them, they will also bring young and fresh Humans, some of them darker than others. Regardless of their hue, for many of them Schools will be sites of torture, death, and trauma *for generations*.[14] Schools will also be sites of Human liberation from their own innovation shackles. They will summon their linear invention of time to manage and regulate what happens in schools, and then, they will engineer yet another concept, **Normal**.[15] After decades of preaching to (or praying on) Humans, by the year 2020 Normal will be canonized. As the new patron saint, Humans will be praying for Normal to bring an end to the devastation. They will turn to The Economy to ask for salvation, and in return they will offer their youngest flesh in sacrifice. They will trade the flesh and bones of those who they will code as from "low SES contexts, jobless households, children with special needs either physical or psychological, children with language other than English backgrounds and refugee populations, rural and remote contexts, and Indigenous Australians and Torres Strait islanders."[16] They will trade them in the name of another fashionable invention, **Equity**. But they will perish.

After ravaging the Humans in the year 2020, entities from the virus world will return with such growing esteem that even St. Normal and The Economy will turn their backs against Humans. They—body-spirit-beyond—will be left to die in the largest sweeping massacre Humans will ever see. Viruses will break into Human supremacy so fast that Humans will have no time to rely on all their failed innovative concepts. All Humans will have left to do is seek asylum in other planes of existence devoid of all the very concepts that forced their displacement, starting with the concepts of self and others. That will be another life beginning.

Bedtime lil one! I'll tell you the rest of the story on another starry night.

★★★★

Three key threads in this story that run through the rest of the volume are *women, planes of existence*, and *futures*.

Women

We could say that it is a coincidence that the chapters for this collection are by women (that is beings historically made up through biocentric scientific fictions as "women," biologized as such to perform the reproduction of—and the built-in resistance against—coloniality and its various rituals). Yet, as Afrofuturist Queer artist Janelle Monáe reminds us in her song Django Jane

We gave you life
We gave you birth
We gave you Earth
We fem the future

The "we" is *Black women*, and also Brown women, and Indigenous women who have been serving as custodians of Country, birthing life, earth, futures, and scholarship. Bo-i' (Chapter 5), the spiritual leader, dreamer, and interpreter of dreams, from Mt. Apo in Mindanao to Turtle Island continues the Obo-Manobo struggle to protect the people-and-land. Sky Woman, in the Oneida Creation story (Chapter 8), gives birth to her daughter, who in dying births life around us—all beings with the spirit of these origins from the celestial realm and without distinction between animate/inanimate. These two chapters written by Grace Simbulan and Oneida Elder Roberta Hill, like the rest of the contributions in this volume, are giving acts of care for Country, land, life and the peoples that are part (not possessors) of it. Hence, it is no coincidence that women scholars responded to the invitation to write for this book, to delve into futures and the various forms through which the continuation of life takes place. Despite much repression, prohibition of scholarship, and even burning of our bodies then *and* now, we, as the *Wati't* in the *Ancestral 2020 vision* above, have been and continue to be the future(s). Bo-i' and her dreams and Sky Woman and the creation do so by mentoring the next generation of dreamers and dream readers, and underpinning the Haudenosaunee Thanksgiving Address "to show people how to deal with grief and death."

Planes of existence

The *ancestral 2020 vision* story also speaks of death and the futures of Humans made viable in seeking refuge in other planes of existence. Death here, as in the Thanksgiving Address, refers to the thousands of lives lost in the COVID-19 pandemic and related systemic dynamics (medical racism, late capitalism, policing, etc.). Death also refers to the end of the world, in the Kichwa story of the pass of eras (Chapter 2), when the Earth will take everything around us and all Human inventions underground, when Humans seek refuge in other planes of existence. Maya K'iche' María Jacinta Xón Riquiac introduces us (in Chapter 3) to the undulating movement that time takes and to planes of existence. According to Xón Riquiac,

> There are several planes. The plane of the *Rajaw q'ij alixak*/the owners of the thirteen groups of twenty days that govern the time of human existence. The plane of the *Q'atab'al tzij*/authorities of the different specialties of the health of the spirit, mind and body; authorities of the different

> manifestations of the beautiful and the tragic; those in charge of delivering justice, among others. The plane of the *Ujer winaq*/the past and the *Qitqa qamam*/grandfathers and grandmothers, generations of human beings for which we exist. The plane of the timeless *Majun ke'unatasaj*/the forgotten, the living or dead beings that no one remembers. The plane of *Ri aj paraqan ja*/those who take care of the beams of the buildings, and the plane of those who oversee the *Tzajalinik*/rounds in the day with a frequency of 12 hours, 6:00 a.m. and 6:00 p.m.

It is here, in the plea to seek refuge in other planes of existence after the human-engineered world has vanished, in what Dharug woman Jo Anne Rey (Chapter 1) calls "extinction industries," that unseen futures under a different set of relations may begin to unfold. At that end of the world where self, other, the individual, difference, human codes, the economy, normal, and equity will have no further use, they will die a much-awaited final death. These human inventions will have no access to any of the planes of existence. Such death will take place, to borrow Jo Anne Rey's words, in "relating ourselves in the Universe differently." Other planes of existence, relationally and simultaneously multiple, will displace Human centrality repositioning ***Country*** as discussed in Jo Anne Rey's chapter, and ***land*** in Jan Hare, Christine Bridge, and Amber Shilling's chapter. Both chapters from Dharug Ngurra (also known as Sydney, Australia) and Turtle Island (specifically Canada and the city of Vancouver, in western terms), offer a set of futures making practices where the colonial institution that is higher education begins *to commit to its own demise*. Higher education does so by *disabling the policies and practices that constantly attempt to impede* the actualization of classes, subjects, experiences, rituals, ceremonies, formations, and acts to displace human power, to reciprocally care for Country, co-create with Country, address Indigenous and Black priorities situated outside human-centric logics, and claim sovereignty.

Futures

Outside of the "Human-making institutions" that are universities and schooling systems in general, Nicole Tu Maung (in Chapter 4) takes us into the Snake Temples of Myanmar. There, the formations of futures that traverse through the understanding that what is Indigenous to these places, spaces, and faith temples is not "the Human", but other-than-human beings that take place, come from place, and are place. Resident *Nat* spirits' novel identities tied to spiritual ancestry are in place attempting to achieve, as Tu Maung asserts, various other futures. These futures, drawing from Sylvia Wynter (in Gioconda Coello's chapter) are harvested by "hybrid-auto-instituting-languaging-storytelling" processes in excess of the Human, homoeconomicus, color coded biocentric being, and normal prototype. Indigenous stories, Gioconda Coello tells us, "do not simply allow the continuation of knowledge but conjure the presence of untold possibilities and relations among people and with land." The

Amazon forest, as the temples of the *Nat* spirits, are sites of uncountable possibilities of futures irreducible to the singularity of policy mandates and regulations. The forest, temples, the dreams, the land, Country, and their signs "house all the possibilities of the past" (Coello) in the lives lived by *Wati't* (in the Ancestral 2020 Vision story), the ancestors that came after her, the *Majun ke'unatasaj*/the forgotten, the living or dead beings that no one remembers, the *Rajaw q'ij alixak*/the owners of the 13 groups of 20 days that govern the time, and many more (Xón Riquiac). An infinite multiplication of futures arrives as ancient people come to live outside again (Kichwa perspective).

Back to schools

Let us now return to schooling as survivance sites of demonstrable Indigenous futurity in spite of it all. At the heart of dislocation, as is common for Indigenous peoples across the globe, loss often concludes the story of colonial cruelty on peoples/ways of life/and forms of life. Ñusta Carraza Ko's father's trajectory produces a curriculum of that which coloniality could not defeat, capture, or conquer. Carranza Ko's (Chapter 6) analysis exposes false starts, dead ends, failures, and detours in the colonial project of schooling still taking place after hundreds of colonial years. Despite consistent colonial attempts to fracture and break, the ongoing relational struggle that Indigenous peoples of the Carranza kin maintain until the present day is a testament to ancient peoples living, outside, on the ground, again. After the Earth has ended so many times, and taken everything underground, so many times, ancient peoples live. Outside. On the ground. Again. COVID has forced Humans (that is those not locked up, in bondage, enslaved, or caged at a border) to retreat to their own places, keeping their young out of schools. The demand has created a new (?) form of dislocation and loss, a loss of a different kind. The resonances with Ñusta's father's story, though indirect, are unmistakable. At a loss, Humans are exasperated about children missing school content, missing out, unlearning skills, and national and international standardized test scores. Here the 2020 pandemic begins to expose the false starts, dead ends, failures, and detours of 21st century colonial schooling, where Humans will need to learn to lose themselves. In what follows, the ancestral 2020 vision continues as we wish to dream well into this undulating moment where we too seek to lose ourselves.

★★★★

Ancestral 2020 vision: Chronicle of a death foretold

Part II

Wati't, will you tell me the rest of that origin story?
The one where another life begins in the future.

Right. The story of a future inhabited by peculiar beings who will call themselves Humans. Thousands of years from now, in the year 2020 entities from the virus world will decimate thousands of them. Humans will be forced to flee. They will mistakenly think of the virus in the only frame they know: War.[17] So in the first months of a rising decade in the 21st century, they will wake up and will feel as if bombs were dropped on them as they have dropped bombs on others. With no time to pack their bags with Human supremacy, and the rest of their useless belongings including belonging itself, they will lose being and becoming Human.[18]

What will those planes of existence look like Wati't?

Full of **wonder** and **wilderness**![19] At first they will be afraid of what they will encounter. With no Normal saint or Economy god to misguide their existence, they will be swept with bewilderment. The first thing they will realize is that they are just one more element of an endless constellation and **not the centre of it**. Some of the cleverest ones, well they will not be that clever, really! But some of them will smuggle their useless concept of the Individual, which will have no currency in other planes. Their existence will be **relational**. Upon entry to these unfamiliar planes of existence, Humans will disappear and become a **collective** of moving particles. **Friction, magnetism, and sonic waves**, not Identities, Borders, or Institutions, will characterize their movements in the new planes.

Remember that genius Human invention they called Schools? Well before being forced by the 2020 virus entity to flee the world as they knew it, Schools were Institutions. In these planes Schools will be **sites of encounter**. They will be located throughout and will not require a building called "school" for learning to take place. Because some habits are hard to die, some of them will invoke obsolete modern ideas to say "*schools are the best place for children to learn.*" But the majority of them having learned the lessons from the 2020 pandemic will ask: **To learn what? To learn how? To learn whose knowledge? Can knowledge be owned? What counts as learning? What learning counts? Where, with what, and with whom can learning take place?** In the new plane of existence, they will realize that moulding children into rubrics, test scores, statistics, international comparative data, and the next capitalist battalion of citizens will bring their demise. It will by 2020 and will again after 2020.

Instead of learning to become settler colonialists, at the site of encounter children will teach and learn ***about*** **colonialism** and the history that forced them to seek refuge in new planes of existence. Instead of taking, children will learn to **borrow and give back** in full respect of the sentient beings they will be in commune with. Children will return to their wilderness without the Human fear of "being backward" or "falling behind." That will be pre-2020 history. Colour codes for children's bodies will no longer be the norm. In fact, there will be no norms, gender or otherwise. There will not even be "children" as the category upon which "adults" manage and attempt to control the/

their future. Through wondering into the histories of human grids and classificatory regimes through online or VR learning, they will regain a sense of **humility**. That will be the ultimate social and emotional learning lesson, resulting in new worlds unfolding. How marvellous!

Maltyox wati't

★★★★

Before the spirit of writing leaves these pages and moves on to worlding elsewhere, we will leave you/us with two vignettes where Brown, Red, and Black futures are taking place. The two vignettes are sites of encounters. They are movements of collectives imbued with **friction, magnetism, and sonic waves**. You will notice that their relational possibilities respond to entities of the viral world (in this case COVID-19) outside Human war logics. In them, we may be witnessing ancient people coming to live outside again, in Kichwa terms, the *Ujertaq Winaq*, in Maya-k'iche' terms, the people of before/The people of always.

Resistances, insurgencies, and struggles for life in extermination times

Via Zoom, the online meeting platform popularized during the pandemic, Maya K'iche scholar Gladys Tzul Tzul speaks of plural and communal life in Guatemala. The platform is a virtual conversation under the name *Political Ecologies of the Pandemics.*[20] Gladys begins with a description of Indigenous life in communities pre-COVID. Prior to the arrival of the virus, Indigenous choreographies in Guatemala were characterized by struggles against land evictions imposed by monocultural agrarian policies by the nation state. Indigenous peoples were protecting water against mining companies. Hundreds of Indigenous men and women were safeguarding the entrance of communities (**sites of encounter**) to stop mining machinery from entering their territories (**friction**). The water ceremonies kept taking place (**site of encounter**). February and March are corn planting season. Indigenous planting life went on, as did the preparations for Indigenous festivities in various communities (**magnetism)**. Meanwhile the legal struggle for treaty ratification that guarantees protections for Indigenous seeds and grains continued. There was a vitality and a functioning communal structure in place (**magnetism**), Gladys explains. Some were more solidified than others as expected. The recent genocide, after all, caused severe damage to these long-lasting forms of communal life. *Life and the reproduction of life are central to these Indigenous communities.* Thus, communal structures are dedicated to care for water sources, the land's fertile periods, and safeguarding communal life overall (**collective**). That is the **relational** continuation of life in place by the time the virus took hold of Human life in Guatemala.

Gladys described this collective magnetism as political plurality through which Indigenous communities have been able to respond and endure the pandemic. *Liderxs comunales* (community leaders), *curanderxs* (healers), and *comadronxs* (midwifes) undertook the task of caring for people via radio messaging in various Indigenous communities (**sonic waves**). How to boost the immune system of women who recently gave birth, what elements to use for personal protection, what plants, trees, and seeds to use to care for the health of all, are among the multiplicity of medical knowledges broadcasted in Mayan languages. The government suppression strategy to marginalize and delegitimize these knowledges, practices, and politics (all elements of **wilderness** and **wonder**) failed. Indigenous leaders, water committees, school committees, among others, organized to create the conditions for Mayan peoples to self-isolate. A key measure in this process was the decentering of the historical Indigenous market days (twice a week) where multiple Indigenous farmers from various communities gather centrally to trade their products. They created multiple points of trade across the various Indigenous territories (**site of encounter**) to guarantee the distribution and supply of food for Indigenous people and those in the city, minimizing the loss of their own labour while protecting everyone's health and wellbeing. They created communal kitchens and *comedores solidarios* to feed many.

Through Zoom, people from around Abya-Yala gather to talk in Kichwa in a language Summer course which would usually take place in Ecuador. "Social distancing" (though physical distancing is a more appropriate term) in the classroom is navigated through virtual closeness in living rooms and bedrooms in different countries. Nely, a chakra mama[21] and Kichwa instructor asked us: "How is the virus doing where you are?" (**site of encounter**) and the question took her interlocutors in many directions. Silence came and she looked pensive. She said that in the *Oriente* (Ecuadorian Amazonia) the virus is in everyone's home. She thought probably everyone had it by now, herself included, even though she did not have any symptoms. In the area where Nely lives, close to Tena, families share their recipes with each other (**collective**). No one wants to go to the hospital out of fear of dying (**friction**). There is no trust that the care needed will be provided at hospitals. It is not luring to be in isolation between white walls, surrounded by overworked and underpaid nurses and doctors away from their network of people, plants, and places that provide care and medicine. The family and extended family support the physical, emotional, and spiritual health of people helping every-body to go through their "virus-ful" time (**relational**). People in the surrounding areas in the *Oriente* have found out how to help the body get through the pains and discomforts the virus brings, Nely explained: Gargles with warm lemon water and salt three times a day; lemongrass tea with lemon; turmeric and ginger tea to complement; daily shower, not in the river, but at home with water boiled with sweet herbs (**borrow**). This approach to wellbeing is not new among Kichwa people in the *Oriente*. The sense of unreliability in public health is not new either. What is new is the relationship with this particular virus and the recipes and procedures they have come up with after spending time with the virus. Rather than seeking imprisonment of people or viruses, this encounter fosters attention. This experience might strengthen communitarian networks of care and a sense of vulnerable listening between people and more-than-human beings. In a virusful time, when all the Human systems and beliefs fail, the sites of encounter between beings of different kinds might allow for visions and enactments of existence otherwise.

What other sites of encounter might emerge of that which is to come?

Notes

1 Sylvia Wynter, "1492: A new world view," in *Race, Discourse, and the Origin of the Americas: A New World View*. Vera Lawrence Hyatt and Rex Nettleford, eds. Washington: Smithsonian Inst Pr, 5–57, 1995.

2 Frantz Fanon, *The Wretched of the Earth*, Reprint edition. New York: Grove Press, 1961.

3 The title and orientation of this piece are inspired by two important works: Munanjahli and South Sea Islander woman Chelsea Bond's poem Dear Ancestor (in Alison Whittaker, *Fire Front: First Nations Poetry and Power Today*. University of Queensland Press, 2020) and Caribbean writer Gabriel García Márquez's novel, *Crónica de una muerte anunciada* (Barcelona: Bruguera, 1981). In an interview in 1981, shortly after the novel was released, Gabo speaks of violence which "in Latin America, and mainly in Colombia, is a phenomenon in all of its history, something that comes from Spain. Violence is the greatest midwife of our history" Jesús Ceberio, "García Márquez: «Crónica de una muerte anunciada» es mi mejor novela," *El País*, April 30, 1981, sec. Cultura, https://elpais.com/diario/1981/05/01/cultura/357516008_850215.html.

4 The beginning of this story has the futuristic sense inscribed in the first words that sound out of the film Black Panther (Ryan Coogler, *Black Panther*, 2018.). *Wati't* is grandmother in Maya-K'iche' and a term that calls for ancestry here and now, for story telling that desires imagination in a time of imaginative scarcity. This is a story of and for making futures.

5 On the emergence of human/ism see Sylvia Wynter's extensive body of work including Sylvia Wynter, "The re-enchantment of Humanism: An interview with Sylvia Wynter," *Small Axe*, 2000.

6 For an account of the COVID-19 pandemic, see https://covid19.who.int/

7 Bawaka Country et al., "Co-becoming Bawaka: Towards a relational understanding of place/space," *Progress in Human Geography* 40 (4): 455–75, 2016. https://doi.org/10.1177/0309132515589437.

8 A key text to grapple with the invention of difference in the context of migration, highly relevant to these pandemic times, is Jacques Derrida, *Of Hospitality*. Stanford: Stanford University Press, 2000.

9 See, for example, Silvia Rivera Cusicanqui, *Un Mundo Ch'ixi Es Posible*. Buenos Aires, Argentina: Tinta Limón, 2014; and Ibram X. Kendi, *Stamped from the Beginning*. Bold Type Books, 2016.

10 On the making up of people, see Ligia López López, *The Making of Indigeneity, Curriculum History, and the Limits of Diversity*. New York: Routledge, 2018.

11 For more on how naming acts on people, see, for example, Ligia (Licho) López López, "Refusing making: Journal of curriculum and pedagogy: Vol 0, No 0," *Journal of Curriculum and Pedagogy*, 2019. https://doi/full/10.1080/15505170.2018.1541828.

12 In capitalism, "the common good is private property and individual representation" Gladys Tzul Tzul, "La forma comunal de la resistencia," *Revista de la Universidad de México*,

April 2019. https://www.revistadelauniversidad.mx/articles/7a052353-5edf-45fe-a7ab-72c6121665b4/la-forma-comunal-de-la-resistencia.

13 Daina Ramey Berry, *The Price for Their Pound of Flesh: The Value of the Enslaved, from Womb to Grave, in the Building of a Nation*. Boston: Beacon Press, 2017. This work is key in understanding this trading dynamic in chattel slavery history.

14 For a visual example of current schooling dynamics in the context of Aboriginality in Australia see, In My Blood It Runs (Maya Newell, *In My Blood It Runs*, Documentary, 2019, https://inmyblooditruns.com/.)

15 For a visual and popular culture challenge to this, see for example, La Red Interuniversitaria y del Caribe de Discapacidad y Derechos Humanos, "Comunicado en solidaridad con Chile," accessed May 6, 2020, http://red-universidadydiscapacidad.org/novedades/item/328-comunicado-en-solidaridad-con-chile.

16 In the beginnings of the COVID-19 pandemic, the Australian government commissioned five research reports focussing on education. This statement comes from one of the reports produced at the University of Melbourne. Janet Clinton, "Supporting vulnerable children in the face of a pandemic: Supporting vulnerable children in the face of a pandemic" (Melbourne: Centre for Program Evaluation. Melbourne Graduate School of Education, April 25, 2020), 4.

17 In reference to the war discourse employed by many country leaders to combat and defeat the virus. Some examples include the 45th president of the United States, the "war-time president", calling to defeat the virus with which "we are at war"; Queen Elizabeth evokes language from WWII ("we will meet again") in an address to the kingdom; and Italy's Prime Minister Conte describes the moment as the "darkest hour" echoing Winston Churchill's words during WWII.

18 For an important history of humanity see, for example, Zakiyyah Iman Jackson, *Becoming Human: Matter and Meaning in an Antiblack World*. NYU Press, 2020.

19 We draw from Jack Halberstam's sense of the wild here. Jack Halberstam, "Wild thing: Zombie Humanism at the end of the world". University of Richmond, 2016.

20 See the full conversation here: https://www.youtube.com/watch?v=o46lkUYhuq0&list=PLpl-iZjcowtIZ6OaweN4-ZRMSI8t27KOh&index=30&t=0s

21 A woman who works the land planting and harvesting, either for self-sustaining or for commercial purposes.

References

Berry, Daina Ramey. *The Price for Their Pound of Flesh: The Value of the Enslaved, from Womb to Grave, in the Building of a Nation*. Boston: Beacon Press, 2017.

Ceberio, Jesús. "García Márquez: «Crónica de una muerte anunciada» es mi mejor novela." El País. April 30, 1981, sec. Cultura. https://elpais.com/diario/1981/05/01/cultura/357516008_850215.html

Clinton, Janet. "Supporting vulnerable children in the face of a pandemic: Supporting vulnerable children in the face of a pandemic." Melbourne: Centre for Program Evaluation. Melbourne Graduate School of Education, April 25, 2020.

Coogler, Ryan. *Black Panther. Super Hero*, 2018. https://www.marvel.com/movies/black-panther

Country, Bawaka, Sarah Wright, Sandie Suchet-Pearson, Kate Lloyd, Laklak Burarrwanga, Ritjilili Ganambarr, Merrkiyawuy Ganambarr-Stubbs, Banbapuy Ganambarr, Djawundil Maymuru, and Jill Sweeney. "Co-Becoming Bawaka: Towards a relational understanding of place/space." *Progress in Human Geography 40*, (4): 455–475, 2016. doi:10.1177/0309132515589437.

Cusicanqui, Silvia Rivera. *Un Mundo Ch'ixi Es Posible*. Buenos Aires, Argentina: Tinta Limón, 2014.

Derrida, Jacques. *Of Hospitality*. Stanford, CA: Stanford University Press, 2000.

Fanon, Frantz. *The Wretched of the Earth*. Reprint edition. New York: Grove Press, 1961.

Halberstam, Jack. "Wild thing: Zombie Humanism at the end of the world." Richmond, UK: University of Richmond, 2016.

Jackson, Zakiyyah Iman. *Becoming Human: Matter and Meaning in an Antiblack World*. New York: NYU Press, 2020.

Kendi, Ibram X. *Stamped from the Beginning*. Reprint edition. Bold Type Books, 2016.

La Red Interuniversitaria y del caribe de Discapacidad y Derechos Humanos. "Comunicado en solidaridad con Chile." Accessed May 6, 2020. http://red-universidadydiscapacidad.org/novedades/item/328-comunicado-en-solidaridad-con-chile

López López, Ligia. *The Making of Indigeneity, Curriculum History, and the Limits of Diversity*. 1st edition. New York: Routledge, 2018.

López López, Ligia (Licho). "Refusing making: Journal of curriculum and pedagogy: Vol 0, No 0." *Journal of Curriculum and Pedagogy*, 2019. https://www-tandfonline-com.ezp.lib.unimelb.edu.au/doi/full/10.1080/15505170.2018.1541828

Márquez, Gabriel García. *Crónica de una muerte anunciada*. Barcelona: Bruguera, 1981.

Newell, Maya. In My Blood It Runs. Documentary, 2019. https://inmyblooditruns.com/

Tzul, Gladys Tzul. "La forma comunal de la resistencia." Revista de la Universidad de México, April 2019. https://www.revistadelauniversidad.mx/articles/7a052353-5edf-45fe-a7ab-72c6121665b4/la-forma-comunal-de-la-resistencia

Whittaker, Alison. *Fire Front: First Nations Poetry and Power Today*. Brisbane: University of Queensland Press, 2020.

Wynter, Sylvia. "1492: A new world view." In *Race, Discourse, and the Origin of the Americas: A New World View*, edited by Vera Lawrence Hyatt and Rex Nettleford, 5–57. Washington: Smithsonian Inst Pr, 1995.

———. "The re-enchantment of Humanism: An interview with Sylvia Wynter." *Small Axe*, (8), 2000.

Index

For Product Safety Concerns and Information please contact our EU representative GPSR@taylorandfrancis.com
Taylor & Francis Verlag GmbH, Kaufingerstraße 24, 80331 München, Germany

www.ingramcontent.com/pod-product-compliance
Ingram Content Group UK Ltd.
Pitfield, Milton Keynes, MK11 3LW, UK
UKHW020005200926
475702UK00015B/2487
* 9 7 8 0 3 6 7 6 7 3 0 3 1 *